AF748821

ENERGY, ENVIRONMENT AND GEOPOLITICS IN EURASIA

This book advances our understanding of security and its intricate interactions with geopolitics and the environment in Eurasia.

Norman A. Graham and Şuhnaz Yılmaz focus on Eurasia, where the energy-water-food nexus has emerged as a vital aspect of political economy and increasingly as a decisive factor for human security. As clearly revealed during the Russian invasion of Ukraine, this nexus rests on a precarious balance. Graham and Yilmaz argue that Central Eurasia is currently "Running on Empty" and highlight the key environmental challenges, including water quantity and quality and food security. The authors draw on their extensive fieldwork in countries including Azerbaijan, China, Georgia, Kazakhstan, the Russian Federation, Türkiye, and Uzbekistan to assess the interests and impact of pivotal actors and evaluate the competition and complementarities of these actors regarding water, energy, food security, and foreign policy imperatives. They also examine the broader interaction and implications of security at multiple levels by analyzing the local, national, and international factors in light of geopolitical and environmental challenges.

Taking a novel and highly interdisciplinary approach, this book will be an important resource for students and scholars of energy and food security, political economy, international conflict and cooperation, and natural resource politics.

Norman A. Graham is a professor at the Center for European, Russian and Eurasian Studies and James Madison College, Michigan State University.

Şuhnaz Yılmaz is a professor of International Relations and serves as the dean of the College of Administrative Sciences and Economics and the director of the Graduate School of Business at Koç University, Istanbul.

Routledge Explorations in Environmental Studies

Coastal Disaster Risk Management in Bangladesh
Vulnerability and Resilience
Edited by Mahbuba Nasreen, Khondoker Mokaddem Hossain and Mohammed Moniruzzaman Khan

A History of Radioecology
Patrick C. Kangas

Satire, Humor, and Environmental Crises
Massih Zekavat and Tabea Scheel

UN Human Rights Institutions and the Environment
Synergies, Challenges, Trajectories
Sumudu Atapattu

Myth and Environmentalism
Arts of Resilience for a Damaged Planet
Edited by Esther Sánchez-Pardo and María Porras Sánchez

Christian Environmentalism and Human Responsibility in the 21st Century
Questions of Stewardship and Accountability
Edited by Katherine M. Quinsey

Energy, Environment and Geopolitics in Eurasia
Search for Security in the Water-Energy-Food Nexus
Norman A. Graham and Şuhnaz Yılmaz

For more information about this series, please visit: www.routledge.com/Routledge-Explorations-in-Environmental-Studies/book-series/REES

ENERGY, ENVIRONMENT AND GEOPOLITICS IN EURASIA

Search for Security in the Water-Energy-Food Nexus

Norman A. Graham and Şuhnaz Yılmaz

Routledge
Taylor & Francis Group
LONDON AND NEW YORK

earthscan
from Routledge

Designed cover image: Getty images

First published 2024
by Routledge
4 Park Square, Milton Park, Abingdon, Oxon OX14 4RN

and by Routledge
605 Third Avenue, New York, NY 10158

Routledge is an imprint of the Taylor & Francis Group, an informa business

© 2024 Norman A. Graham and Şuhnaz Yılmaz

The right of Norman A. Graham and Şuhnaz Yılmaz to be identified as authors of this work has been asserted in accordance with sections 77 and 78 of the Copyright, Designs and Patents Act 1988.

All rights reserved. No part of this book may be reprinted or reproduced or utilised in any form or by any electronic, mechanical, or other means, now known or hereafter invented, including photocopying and recording, or in any information storage or retrieval system, without permission in writing from the publishers.

Trademark notice: Product or corporate names may be trademarks or registered trademarks, and are used only for identification and explanation without intent to infringe.

British Library Cataloguing-in-Publication Data
A catalogue record for this book is available from the British Library

ISBN: 978-1-032-44749-0 (hbk)
ISBN: 978-1-032-44750-6 (pbk)
ISBN: 978-1-003-37373-5 (ebk)

DOI: 10.4324/9781003373735

Typeset in Times New Roman
by codeMantra

CONTENTS

ACKNOWLEDGMENTS

This volume began with a series of conversations between the authors during summer education abroad encounters between the students and faculty of James Madison College of Public and International Affairs at Michigan State University and those of Koç University's international relations faculty in Istanbul, Türkiye. As professors of international relations from different continents, we marveled at the changing geopolitical environment facing Europe and Eurasia, while the Putin administration took hold in Russia. Some of our collaboration began to explore the extent to which complex water, energy, and food interactions were becoming central to the structure of the emerging international system and the challenges facing the newly independent successor states to the Soviet Union. We first owe a debt of gratitude to the numerous students from both institutions over many years of this study abroad experience who raised questions and challenged our initial interpretations and projections.

As our ideas began to take shape, we benefited from a series of visits to various sites across Europe, Russia, and Eurasia, especially to the EU and NATO headquarters in Brussels, and the capital cities of nearly all of the post-Soviet states and the former Warsaw Treaty Organization members, most notably the Czech Republic, Hungary, Poland, and Romania. Madison College's International Relations in Brussels Summer Study Program with the Université Libre de Bruxelles and its Economic and Political Transition in Eastern Europe Summer Program were important venues for these encounters. We also greatly benefited from our visit to and presentations at the China University of Political Science and Law and the Chinese Academy of Social Sciences. More specifically, we would like to acknowledge the dedicated research assistance over the years provided by Cody Schulz, Nora K. Graham, Allie Virginski, Kate Wills, Isabelle Borr, Jake Maly, and Rachael Lee at Michigan State University, and Duygu Sever Mehmetoğlu, Ariel Gonzalez

Levaggi, Tolunay Öndül, Derya Çalış, Deniz Tokmakoğlu, Gülşen Doğan, Zehra Yağmur Yılmaz, and Özge Karsu at Koç University.

A large group of MSU and KU faculty and visiting scholars have been important to us: inspiring this project, guiding our questions, and commenting on our initial answers. These would include Abigail Bennett, George Bird, Jiquan Chen, Emine Evered, Kyle Evered, Sergey Fedunyak, Eric Freedman, Anthony Olcott, Martha Brill Olcott, Matt Pauly, Michael Polyak, Jiaguo Qi, Simei Qing, Michael G. Schechter, Uli Schamioglu, Lawrence Scott Sheets, Jamie P. Shea, Matt Zierler, and M. Richard Zinman. At Koç University, we would like to acknowledge Belgin San Akça, Ziya Öniş, Bahar Rumelili, Caner Bakır, Ayşegül Özsomer, and Dilek Barlas for their valuable insights and support.

We owe also a collective debt to our inspiring teachers of Russian and Eurasian Studies, Middle Eastern Studies, and International Relations, especially Zbigniew Brzezinski, David J. Louscher, Donald J. Puchala, Glenda Rosenthal, Marshall D. Shulman, Steve Kotkin, Şükrü Hanioğlu, Şevket Pamuk, and John Waterbury, and to those who helped to encourage our interest in environmental sustainability, most notably Robert W. Cox, William T. R. Fox, Wilfrid L. Kohl, and William Taylor. Three colleagues at Michigan State read and commented on early drafts of several chapters: Folke Lindahl, Steven G. Pueppke, and Volodymyr Tarabara. Our joint discussions with them on Russian foreign policy, European and Eurasian politics, and environmental sustainability have been invaluable.

We both have enjoyed the support of our College and University administrations. Key colleagues and supervisors at MSU include: Deans Sherman W. Garnett, Linda Racioppi, Cameron Thies, and Steven Hanson, and Associate Dean for Research Robert Brathwaite, and International Relations Chair Kirstin Brathwaite. MSU's Center for European, Russian and Eurasian Studies and its Director, Volodymyr Tarabara, the Asian Studies Center and its Director, Siddharth Chandra, and MSU's Asia Hub and its Director, Jiaguo Qi, have been very supportive intellectually and financially. At Koç University, former Dean of College of Administrative Sciences and Economics (CASE) and Vice President for Academic Affairs Zeynep Gürhan Canlı and Associate Dean Nilüfer Aydınoğlu have been a great source of support.

Our international collaborations and interactions with colleagues in leading academic institutions have also been an invaluable source of inspiration and created precious insights and feedback. We are particularly indebted to Robert Crews at Stanford, Alex Cooley at the Harriman Institute, Columbia, and JR de Shazo at the Luskin School of Public Policy at UCLA (who currently serves as the dean of the LBJ School of Public Affairs, the University of Texas at Austin). EU Horizon 2020-FEUTURE project, TÜBİTAK 2219 and Koç University Tüpraş Energy Center (KÜTEM) provided valuable financial support for research. We would like to acknowledge Chris Alden, Buğra Süsler, and Yaprak Gürsoy within the framework of highly inspiring Koç-LSE collaborations and LSE IDEAS program.

We also acknowledge the long-term impact of our research and education partners in Central Eurasia, most notably Sabyr Nurtazin, Talgarbay Konysbaev, and Margulan Iklasov of the Faculty of Biology and Biotechnology, Al-Farabi Kazakh National University in Almaty, Alim Pulatov of Tashkent Institute of Irrigation and Agricultural Mechanization Engineers, Bahodir Eschanov of Tashkent State Agricultural University, and Charles H. Fairbanks, Jr. and Ghia Nodia of Ilya State University in the Republic of Georgia.

Editorial support and peer review comments provided by Routledge were very important to the development and revision process. Jyotsna Gurung and Annabelle Harris were thoroughly professional and responsive to our queries.

Finally, but certainly not least, is the debt we owe to our supportive families: Dr. Anna Kirkwood Graham, Carolyn Donna Kirkwood Graham, Nora Karin Kirkwood Graham. They accompanied us on numerous study ventures to Europe and Eurasia and helped to provide perspective that may well have escaped us otherwise. Prof. Burak Özbağcı, Eda Naz Özbağcı, and Arda Osman Özbağcı, as well as the late Şuhay Yılmaz, Celalettin Yılmaz, and Zeynep Çolak were also a great source of support and inspiration during the long academic journey which culminated in the writing of this book.

1

INTRODUCTION AND THEORETICAL FRAMEWORK

There is growing evidence that the natural world and human activity on earth are out of balance. Global energy consumption and carbon emissions clearly contribute to climate change, impacting water quantity and quality and food security. A disrupted global economic system and concerns about the growing inability of international organizations to respond to trade and territorial disputes undoubtedly complicate our ability to address this imbalance.[1]

All of this has been accentuated by the devastating effects of the COVID-19 pandemic. Moreover, the Russian invasion of Ukraine has also brought energy as well as food security concerns once again into the limelight. Climate change, the pandemic, and regional conflicts have added extensive pressure on the water-energy-food nexus in several respects, and they alert us to the fact that serious disruptions can come about suddenly, perhaps unexpectedly. Thus, we posit that the unprecedented levels of uncertainty in the international arena and the drastic momentum of change compel us to reconceptualize existing paradigms connecting energy and water politics to international and human security. It is important to formulate more comprehensive theoretical frameworks, foreign policy strategies, and domestic policies that integrate energy, water, and food security with sustainability.

In our view, Central Eurasia is now "Running on Empty," in that water quantity and quality are increasingly constrained; energy resource exports provide less reliable income; and there are mounting concerns regarding food security. Moreover, extensive engagement of external actors, especially Russia and the EU, but Europe, the United States, and Turkey, also presents geopolitical challenges and opportunities.[2] The successor states to the USSR in Central Asia and the South Caucasus are transforming markedly, but they are also facing complex domestic and international challenges. This book analyzes Central Eurasia's regional dynamics, as well as its relations with Europe, Russia, China, and other external actors, as Russia

DOI: 10.4324/9781003373735-1

seeks to regain dominant influence over former Soviet regions in its "near abroad," the EU strives to ensure its energy security, China's Belt and Road Initiative plays out, and Turkey acts more assertively beyond its borders as some have termed Pan-Turkism or "Neo-Ottomanism."[3] The uncertainty of all these dynamics is dramatically complicated with the impacts of climate change, the COVID-19 pandemic and its potential successors, and regional conflicts and wars.

Some particularly critical trends in global energy include highly volatile oil prices in global markets; the rising share of LNG in the inter-regional energy trade; the implications of the shale gas revolution turning the United States from a net-importer to a net-exporter; the continued strength in Asian fossil fuel consumption, especially in China; and the technological advances and unprecedented growth of renewables. There are also pressing implications of climate change, depletion and degradation of water resources, and food insecurity. These global trends present numerous challenges, as well as new opportunities for key actors in critical regions such as Eurasia to link vital resources, fossil fuels, and water. They continue to determine the future of extremely volatile regions of Eurasia with major geopolitical implications for regional and global powers. This book analyzes the strategic, political, socio-economic, and environmental impacts of domestic and transnational utilization and governance of these strategic resources. The study also explores the broader implications for the water-energy-food nexus particularly in Central Asia as countries seek to diversify their economies away from dependence on fossil fuel exports (and imports) and the cotton monoculture. Efforts to develop productive and sustainable agriculture within the constraints and imperatives of climate change are an important part of this picture, as are competing claims to water resources that are often of transboundary nature. The fear of over-dependence on sources of energy suppliers which may be unstable or malevolent also remains high on the policy agenda. These concerns have been accentuated with the Russian invasion of Ukraine in February 2022, which threatens European energy security and cannot help but have an impact on the "energy juggernaut" that Michael Klare once exclaimed was the key characteristic and perhaps strength of Putin's economic recovery strategy after 2000.[4]

In this context, there is a need to reassess the prevailing security paradigms. Along with the dominant hard-security concerns which reflect the realities of a conflict-laden region, soft security issues including the economic and environmental aspects of security should also receive more attention. Are there prospects for serious conflict over scarce resources, especially water? Is the continuing dependence of European states on Russia for oil and gas supplies increasingly problematic, as clearly revealed during the Russian invasion of Ukraine? Is food security likely to affect populations in high poverty and arid regions to a degree that elevates this concern to the level of the usual conceptions of national security?

At the theoretical level, this book posits that the definition of national interest in geopolitical terms and approaching it often as a zero-sum game should be accompanied by a broader discussion of economic dynamics and institutional mechanisms. Hence, in this study, the dominant realist and neo-realist approaches are enriched with a more diverse theoretical discussion, including liberal institutionalist interpretations. A comprehensive framework should take into account the changing international context and geopolitical considerations, strategy and tactics, domestic political and economic influences, and the role of leadership. This, of course, requires a systematic assessment of differences of interpretation in terms of their relative weight. We posit that the changing international context, geopolitical factors, and their interaction with domestic political and economic influences against a backdrop of regional conflicts and increasing environmental challenges shape the complex dynamics of the water-energy-food nexus.

Hard security issues have dominated the agenda of the Eurasian powers. While these remain at the core of regional geopolitics, broader security issues, particularly different dimensions of energy, water, and food security, also need more emphasis and attention. The International Energy Agency (IEA) defines energy security as "the uninterrupted availability of energy sources at an affordable price."[5] The IEA also highlights the multidimensional nature of energy security by asserting that while "long-term energy security mainly deals with timely investments to supply energy in line with economic developments and environmental needs," short-term energy security needs are defined by "the ability of the energy system to react promptly to sudden changes in the supply-demand balance."[6] By also stressing environmental concerns, the EU Commission has set out a plan of action for a *European Green Deal*

> to make the EU's economy sustainable by turning climate and environmental challenges into opportunities across all areas of policy in a way that is fair and inclusive. The European Green Deal is the new growth strategy for an EU economy that is sustainable, cleaner, safer, and healthier.[7]

Russia's *National Security Strategy to 2020* document revealed how energy producers approach energy security. The document stated that increasing energy security is possible by creating new markets for energy resources, developing energy-saving technologies, reaching sufficient standard quality for energy sources, and ensuring stable functioning of the system of energy and fuel provisions.[8] Overall, as indicated by the Ukraine case, there is little doubt that Russia has used energy as a tool of foreign policy and strives to weaponize it to facilitate its aggressive regional moves.[9]

Geopolitical factors also shape energy interactions among key actors.[10] Hence, energy security continues to dominate the agenda of regional, as well as global powers.[11] As stressed by Jason Bardoff, "transforming an industry that has defined the modern era will have profound consequences on the global order," yet this transformation's geopolitical implications could be "subtle, complex, and counterintuitive."[12]

Water security has also emerged as an indispensable part of national and regional security calculations. The quantity and quality of water (given climate change and increasing glacial melt, the continuing challenge of inefficient irrigation systems for unsustainable agriculture, and the pollution from ineffectively regulated industrial, agricultural, and urban sewage systems) pose serious challenges to many Eurasian governments. This frequently serves to increase tensions between the governments and populations of Central Eurasia, as river systems cross borders without robust international agreements.[13] As early as 2009, the World Economic Forum defined water security as "the gossamer that links together the web of food, energy, climate, economic growth and human security challenges that the world economy faces over the next two decades."[14] Comparably, there are other specialists, who emphasize the environmental dimension of the term more and define water security as "sustainable access on a watershed basis to adequate quantities of water, of acceptable quality, to ensure human and ecosystem health."[15] As water issues are politically intertwined with the economy and the environment, the efficient use of this vital resource is a priority for most states.

The common traditionalist school of security views military power as the ultimate source of state power. Problems with energy, water, and food security are seen as "soft" security issues and therefore less significant.[16] We argue that one must broaden the concept of security to embrace political, economic, social, and environmental aspects, in addition to questions of military power.[17] The effect of environmental change and resource scarcity is increasingly viewed as a strong challenge to states and societies, and has potential to lead to economic decline, social turmoil, border disputes, and large scale migration.[18] A political ecology approach stresses the interactions between politics, economics, technology, and the environment and the linkages between natural resources, governance, and power. This theoretical perspective emphasizes power relations, which define access to natural resources. The injustices and disparities regarding the distribution of costs and benefits are particularly emphasized.[19] Dalby presents a compelling case for rethinking the role of geopolitics in evaluating climate security in the Anthropocene.[20] Energy security assessment must analyze actors, processes, and implications regarding the securitization of climate change in comparative perspective.[21]

There are numerous separate treatments of European, Russian, Middle Eastern, and Chinese water, energy, or food policies and politics, but none that provide a systematic comparison of these cases and their interactions in Eurasia. In this context, George G. Eberling's *China's Bilateral Relations with Its Principal Oil Suppliers*, Philip Andrews-Speed and Roland Dannreuther's *China, Oil and Global Politics*, Anatole Boute's *Energy Security along the New Silk Road*, Jim Krane's *Energy Kingdoms: Oil and Political Survival in the Persian Gulf,* and Rajan Menon et al, *Russia, the Caucasus, and Central Asia: The 21st Century Security Environment* present valuable contributions.[22] There are also more general treatments of bilateral relations that examine different aspects of energy security issues in the region, including Paul J. Bolt and Sharyl N. Cross' *China, Russia and Twenty-First Century Global Geopolitics*, Michal Lubina's *Russia and China: A Political Marriage of Convenience Stable and Successful,* and Ali Askerov's edited volume on

Contemporary Russo-Turkish Relations: From Crisis to Cooperation.[23] Caroline Kuzemko's *The Energy Security and Climate Nexus: Institutional Change in the UK and Beyond* presents a comprehensive assessment from a political economy perspective integrating energy security with climate challenges.[24]

Currently, there is a growing literature linking energy security with key issues and concepts in international studies. In a recent study, Van de Graaf and Colgan present a comprehensive framework not only for exploring how oil politics affect international order in terms of power dynamics and systems change but also for analyzing international governing arrangements regarding climate change.[25] Some authors have examined energy security broadly with inescapable attention to Eurasia. For instance, in *The Quest*, Daniel Yergin explores how energy is linked to security and triggers global political and economic change and conflict. He also discusses new complex factors shaping our energy future in *The New Map: Energy, Climate and the Clash of Nations.*[26] In *The New Geopolitics of Natural Gas,* Agnia Grigas stressed the increasing role of natural gas in shaping geopolitics.[27] Adam Stulberg's *Well-Oiled Diplomacy: Strategic Manipulation and Russia's Energy Statecraft in Eurasia* presents a compelling account of Russia's prevalent role in this region shaped by energy politics.[28] Other scholars have focused on energy development more narrowly in the Caspian Basin, touching on security issues, but paying little or no attention to the water-energy-food nexus or indeed of the complicated dynamics of broader regional politics and policies. These engaging studies include Steve Levine's *The Oil and the Glory: The Pursuit of Empire and Fortune on the Caspian Sea* and James Marriott and Mike Minio-Paluello's, *The Oil Road: Journeys from the Caspian Sea to the City of London.*[29] There are few studies thus far that are systematic investigations of water security in Eurasia. There are insightful global assessments like Brahma Chellaney's *Water, Peace, and War: Confronting the Global Water Crisis* or Walter Leal Filho and Dinesh Surroop's edited volume entitled *The Nexus: Energy, Environment and Climate Change* that raise important questions about trends in water security and environmental issues,[30] but water studies on Eurasia tend to be relatively narrow, such as the many assessments of the Aral Sea crisis, without much direct examination of the energy and food connections.

On the nexus more generally, there are a few relevant studies. Corey Johnson et al, *Want, Waste or War?: The Global Resource Nexus and the Struggle for Land, Energy, Food, Water & Minerals* (2014) and P. Högselius (2018). *Energy and Geopolitics,* provide valuable insights. A recent edited volume by Piotr Dutkiewicz, et al., *Eurasia on Edge: Managing Complexity*, also presents an interesting analysis.[31]

In this volume, we undertake to help fill an apparent gap in the literature by bringing together data and analysis on the complex interactions of the Water-Energy-Food Nexus, climate change impacts, and the evolving interactions of regional geopolitics. Current trends make the energy-water nexus and its food security implications particularly important for volatile Central Eurasia and its neighbors. Rich energy resources do not compensate for water scarcity and environmental degradation.

Moreover, Russia's war on Ukraine has exacerbated regional challenges regarding food security. Indeed, the region is in danger of "running on empty" as it seeks to adjust to an increasingly disrupted global economy while searching for security, independence, and broad-based prosperity. The alarming warnings of the leadership of the World Food Programme on the severe cuts in Ukraine (and Russian) wheat, fertilizer, and vegetable oil exports should give us pause. In response to these rapidly changing global and regional dynamics, the key puzzle we explore is how the relevant actors are assessing their national interests, defining their security, and responding to the new challenges and opportunities that are facing them with respect to the water-energy-food nexus. What are the aspects of continuity and change? What impact will natural resource constraints and struggles have on the future dynamics of their relationships? In which instances and under what conditions will these intricate ties affect food security?

Both the Caspian Basin and Central Asia constitute vital areas in the complex chessboard of international politics, particularly due to the rich energy reserves. Our framework for analyzing these questions involves: (i) assessing the interests and impact of pivotal political and economic actors; (ii) evaluating the competition and complementarities of these actors regarding water, energy and food security and foreign policy imperatives; (iii) analyzing the broader interaction and implications of security at multiple levels by examining the local, national, and international factors in light of geopolitical and environmental challenges; and (iv) exploring the understudied link between energy, the environment, and geopolitics.

The water-energy-food nexus has an impact on several dimensions. The first component of the nexus is security. In a context in which the demands for energy and water are high and the resources are scarce, both energy and water appear as crucial to national security. Secondly, water and energy security are crucial for a state's economy, given the drive to reach energy and water resources at an affordable price without major disruptions.[32] Considering the fluctuating prices of energy, high dependence on foreign resources is clearly a threat to national economies. Accordingly, diversification of suppliers and alternative resources through renewables are critical for mitigating economic risks. In the case of water, in terms of energy production, irrigation, industrial, and domestic use, there is always a strong economic component as well. A third important sphere that is closely connected to natural resource security is the environment. A significant dimension of natural resource security is the preservation of natural resources and their utilization with a minimal negative impact on climate change, through serious attention to reducing carbon emissions. Likewise, the main concern of water security is to fight against water pollution and unsustainable use. There is, thus, a need for an increasing emphasis on efficiency in energy and water distribution and usage.

We analyze the relationships between the energy-water nexus and security, the economy, and the environment in an interactive triangular fashion, as depicted in Figure 1.1. In this context, water and energy security are at the core and there

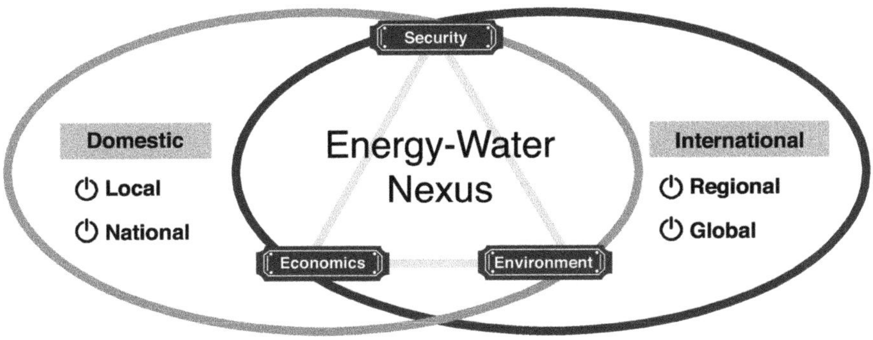

FIGURE 1.1 Energy-Water Nexus: Security, Economics and Environment (SEE).

are two-way relations with the energy-water nexus and the corners of the triangle (SEE), which are security, economy, and environment:

These three pillars are intricately linked to each other. Thus, the overall conception is a hub-and-spoke shape, in which all components are highly interconnected. There is geopolitical, socio-economic, as well as an environmental urgency in the energy and water nexus, both on a regional and the global scale. Moreover, in the domestic context, there are critical implications at the local and national levels. There are also complex interactions between the domestic and international drivers and policy choices. Hence, this tangled web of relations necessitates further research at the empirical and theoretical levels. The book will examine the specific foreign policy frameworks of critical global and regional actors with respect to the geopolitical, economic, and environmental aspects of Eurasian energy and water challenges. It will do so by utilizing a multidisciplinary approach with significant comparative insights.

Water, energy, and food security are generally assessed at the state level. However, a state-centric perspective leads one to neglect the human security dimension. As early as 1994, the UNDP Human Development Report defined human security in two dimensions. Human security, first, means safety from chronic threats like hunger, disease, and repression. It also means protection from sudden and agonizing disruptions in the patterns of life. These threats to human security can be in economic security, food security, health security, environmental security, etc.[33] Thus, natural resource security and human security are strongly linked. Consequently, while analyzing the water-energy-food nexus, the human security implications are critical.

A focus on climate change and the growing imbalance between the natural world and human activity (population growth, urbanization, continuing fossil fuel appetite, inability of governments and societies to achieve broad-based prosperity, and disrupted international economic system, etc.) may be the answer to the increasing challenges. The Gates Foundation argued that Climate Change may well ultimately cause more deaths in the 21st Century than pandemics like the one we faced with COVID-19; but one could also argue that if COVID-19 had been a bit more infectious or the vaccines were a bit slower in arriving, we might be facing something much more devastating. Indeed, Doug Gurr (Director of the Natural History Museum in the United Kingdom) recently warned of the 6th mass extinction.[34] Perhaps we are already in the midst of such a disaster, if we do not rebalance the natural and human world.

Can we connect what we know about the water-energy-food nexus and the failure of governments in Eurasia (and elsewhere) to promote a more sustainable and equitable development process/plan? There is renewed attention to climate change. The Biden Administration has recommitted to the Paris Climate Accord, and the G7 undertook to tackle the twin crises of Climate Change and Biodiversity Loss, as spelled out in a joint communique at its London meeting 21 May 2021.[35]

Confronted with the impact of war, pandemic, and global economic disruption, the United Nations Climate Change Conference (COP 27) held in Sharm El-Sheikh, Egypt in November of 2022 stressed:

> ...that the increasingly complex and challenging global geopolitical situation and its impact on the energy, food and economic situations, as well as the additional challenges associated with the socioeconomic recovery from the coronavirus disease 2019 pandemic, should not be used as a pretext for backtracking, backsliding or de-prioritizing climate action.[36]

COP 27 did make the historic decision to establish a fund for responding to loss and damage from climate change and was fairly strong in not giving in to pressure from national delegations to moderate their previous commitments to reduce carbon emissions in light of the impact of global economic disruption. However, the implementation of actual financial transfers to the fund and the lack of enforcement for national compliance to meeting carbon emission targets were lacking.

One must note that Azerbaijan, Kazakhstan, and Turkmenistan have relied heavily on and benefitted substantially from oil and natural gas exports through much of their histories as independent states, after the unraveling of the Soviet Union. This led to considerable corruption at times, as well as to more general aspects of "the natural resource course." But the rulers of Azerbaijan and Kazakhstan also substantially modernized their economies and improved antiquated infrastructure, particularly in key cities, and invested in transportation and pipeline networks (e.g., the BTC pipeline from Baku, Azerbaijan to Ceyhan on the Aegean coast of Turkey, as well as the subsequent Kazakh and Turkmenistan pipelines to China). However, in many instances, the infrastructure modernization efforts have been scaled back, and more serious efforts to diversify their economies are under way. In Kazakhstan, the authorities have been devoting more attention to upgrading the educational systems, particularly the network of national universities (including the development of Nazarbayev University and current plans to modernize teacher education with substantial foreign assistance) and to expanding and diversifying agricultural production. Nonetheless, the "Bloody January" unrest that erupted in 2022 stimulated by sharp and sudden increases in the price of natural gas pointed to serious discontent with the government and the willingness of Russia to come to the aid of the embattled President, Kassym-Jomart Tokayev.[37]

Similar educational reform and agricultural diversification efforts are currently underway in Uzbekistan, with thus far less reliance on Moscow and some strengthening in its relations with China, Europe, and the United States. Since coming to power in December 2016, President Shavkat Mirziyoyev has seemingly moved on from some of the key policy determinants long held by President Islam Karimov.[38] Diversifying away from unsustainable dependence on cotton as Uzbekistan's key agricultural export product has attracted new commitment; substantial investment in less water intensive fruit and vegetable production is evident, and the post-Soviet collapse of inland fisheries has been addressed in part with a rather successful effort to expand aquaculture infrastructure and commercialization.[39]

It should be stressed, however, that these economic reform efforts are in the face of a challenging set of domestic and international constraints. Striking among these are the environmental challenges that arise from water quantity and quality constraints. There are other regions of the world, like parts of sub-Saharan Africa, that face the threat of desertification more directly and immediately than Central Eurasia, perhaps, but the complexity of the climate change induced and man-made domestic and/or geopolitical constraints in this region is dramatic. How do China, Russia, and Europe fit into this picture? China needs to export energy intensive manufactured products to thrive; and its thirst for fossil fuels seems unabated despite stated public commitment to reduce coal consumption and expand renewables. The demographic and economic imperatives seem both frightening and irresistible. China's Belt and Road Initiative (BRI), initiated by President Xi Jinping in 2013, has rolled out substantially with massive transportation infrastructure investment in China and throughout much of Central Eurasia to reconstruct land and maritime connections between China, much of the rest of Asia, Europe, and the Middle East. One only must observe the highway and high-speed railway construction underway in Western Xinjiang province into Kazakhstan and beyond to see the current and prospective impact. Central Eurasians are both enthusiastic and wary about the economic development opportunities and relationships.[40]

Russia has been striving to coordinate President Putin's Eurasian Economic Union (EaEU—Armenia, Belarus, Kazakhstan, Kyrgyzstan, and the Russian Federation) with the BRI. But the EaEU is also a competitive strategy for pursuing Russian influence in the region. As of now, however, it cannot be regarded as an engine for growth and development in the region and is in danger of being dwarfed by the impact of the BRI and somewhat disrupted by the Russian invasion of Ukraine.

Russia has no real strategy evident for economic prosperity beyond continued exports of fossil fuels and other natural resources. Russia's efforts to respond to Western sanctions after the annexation of Crimea are instructive here.

This included swift action to conclude a long stagnant set of negotiations over Chinese purchase of fossil fuels from Russia, and Putin's declaration of an embargo of agricultural imports from the West. Both actions had important implications for Central Eurasian economic development and environmental sustainability. Unfortunately, the Russian invasion of Ukraine has inserted a chill on this, as regional leaders not surprisingly are cautious about their relationships now with both Russia and China, not to mention the West. One can see this readily in the public response to the initial UN General Assembly resolution harshly criticizing the Russian military operations. Only Georgia and Moldova voted for the resolution, but nearly all the rest in the region, including China, abstained rather than to join Russia, Belarus, North Korea, Syria, and Eritrea as the only states in opposition to the condemnation.

President Putin was worried about a possible decline in European oil and natural gas imports and sought to replace the trade with an increase in exports to China. China got the price deal it wanted and a potentially long-term reliable supplier to meet its growing fossil fuel appetite. Putin's embargo on European luxury goods and agricultural imports was justified not only as a direct response to European sanctions but also as a stimulus to Russian agricultural self-sufficiency. This has not worked out well so far. What it did seem to do was to stimulate increased imports of agricultural products from Central Eurasia.

Putin's reluctance or inability to respond to the emerging global concerns about the impact of carbon emissions on climate change remains unabashed. The increased melting of the glaciers in the Tien Shan mountains of Western China and the mountainous regions of Tajikistan and Kyrgyzstan is clearly a worry to the agricultural interests of Kazakhstan and Uzbekistan; this is particularly troubling, as the latter countries seek to diversify their agricultural production to incorporate more sustainable water use, hoping to learn lessons from the Aral Sea disaster. But the water-energy-food nexus is also evident here in the tensions over cross border water sharing that remain because upstream countries (with little or no fossil fuel reserves) seek to construct dams (viz. Tajikistan's massive Rogun Dam) to generate electricity and divert water efficiently for their own agricultural uses. The old Soviet mechanisms for (enforced) sharing of water and energy among the Central Asian republics have not yet been replaced with an effective or reliable regional agreement among the successor states.[41]

The European Union for many years talked a good game about sustainability and concern for climate change, but the use of heavily polluting types of coal in several member states (e.g., Poland and Germany) and the inability (or unwillingness) of most to reduce dependence on Russian fossil fuel imports seemed to betray real commitment.[42] This has changed, at least for the time being, thanks to Putin's invasion of Ukraine and the subsequent brutal campaign of bombing civilian infrastructure and residential areas. The European Green Deal might yet emerge as a transformative initiative in this respect.

In the meantime, Europe and Russia are at odds over Ukraine, Belarus, and the rise of repression. Indeed, Western concern about the persistence of authoritarianism and the fragility of democracy throughout most of Eurasia tends to hamper productive economic relations between the West and Eurasia.[43] This in turn may clear the way for the growing influence of China and the continued presence of Russia in the region. Kishore Mahbubani cites this as an important element in the economic power of China in the region, and why the Western model of democracy and market economy is not seen as the only viable model for success in Eurasia.[44]

Domestic and international imperatives and constraints are crucial for investigation. In this volume, we offer a multilevel approach to the study of the water-energy-food nexus, which on the one hand analyzes domestic factors at the local and national levels, and on the other hand, explores their implications at the international level by situating them within a regional and global context. In exploring the ultimate interactions between the domestic and international levels, Robert Putnam's two-level game analysis approach provides valuable insights for this effort. Putnam's approach combines both international and domestic constraints in the international cooperation process and presents a compelling theoretical framework for international bargaining.[45] Accordingly, the decision-makers play their games simultaneously on two game boards, namely the international arena and the domestic one.[46] Only when policymakers make a move that is accepted in two levels can international cooperation and its domestic implementation be possible. In this respect, international cooperation for natural resource security needs to consider the domestic game as well.

Fossil fuels remain the dominant component of the global energy mix (comprising approximately 85% of the total demand), while currently, oil persists as the leading fuel. The global energy trends also highlight the increasing role of natural gas and renewables. Oil reserves are heavily concentrated in some of the most volatile regions of the world—the Middle East, North Africa, and the Caspian Sea basin, while gas reserves are located predominantly in the Middle East, the Russian Federation, and Post-Soviet Eurasian countries. According to the US Energy Information Administration, world energy consumption is projected to grow by nearly 50% between 2018 and 2050.[47] This situation will give way to a significant increase in inter-regional energy trade providing the producers and the transporters of the Middle East and Gulf region, as well as Eurasia, with critical geopolitical leverage. Hence, due to increasing interregional trade, the interests and the stakes for both global powers such as Russia, the EU, the United States, and China, as well as regional actors such as Kazakhstan, Turkmenistan, Azerbaijan, Turkey, and Iran will be higher.

The scarcity, rapid depletion, and pollution of water also pose a set of vital issues and a deep-rooted multidimensional problem in Eurasia. Semi-arid and dry conditions affect as much as 79% of this region. What further complicates the scarcity of water in this region are the problems of uneven distribution and the transboundary nature of water resources. In a highly volatile area like Eurasia, where conflicts are rife and trust is low, water is a strategic resource serving as an important factor influencing the domestic and external policies of the major actors in the region. Global warming and climate change serve as a source of additional pressure exacerbating the already existing problems and tensions.

While most studies of hydro politics focus on specific river basins and the interactions between key states involved in the conflict,[48] the literature examining multiple levels—local and international—remains quite limited. This volume seeks to bring in the local as well as the regional dimension. Moreover, it seeks to integrate broader literature examining the links between climate change and international security to our understanding of water constraints and challenges.[49]

Water security is increasingly important because of climate change, pollution, and the transboundary nature of water sources. These dynamics directly affect economic and bargaining relationships between states. Climate change and pollution may turn water-rich countries into water-poor ones, which make them more dependent on outside sources. The Caspian Sea is one of the most polluted habitats in the world. Thus, surrounding countries such as Azerbaijan, Kazakhstan, and Turkmenistan otherwise depend on transboundary rivers, which makes the issue political. According to UNESCO, "To achieve water security, we must protect vulnerable water systems, mitigate the impacts of water-related hazards such as floods and droughts, safeguard access to water functions and services and manage water resources in an integrated and equitable manner."[50]

Food security has also emerged as a dramatic focus of post-Cold War international relations, particularly in the transition and developing regions of Eurasia and Africa. The World Economic Forum drew attention to food security in 2020, noting that by 2050, there will be 9.8 billion people to feed.[51] Indeed, population growth, increased meat-based diets, increased drought, glacier melt and desertification, and regional conflicts complicate already stressed agricultural productivity and distribution challenges in much of arid Eurasia.

Water, energy, and food security are highly interrelated concerns. Hence, one of the significant contributions of this volume is to offer a more comprehensive integrated framework for natural resource security. Global and local markets rely on the availability of natural resources like water, energy, minerals, fish stocks, biodiversity, and other renewable and non-renewable resources to meet the rising consumption. However, the availability of these resources is not assured. Because of this, natural resource security is crucial for the national security as well.[52] An initial depiction of the complexity of some of these interrelationships and their interactive nature is presented in Figure 1.2.

For a comprehensive understanding of the complex security, economic, and environmental aspects of water-energy-water nexus, we have employed a mixed-method approach. For capturing broader patterns over time with respect to international conflict and cooperation, we drew data from standard international databases related to energy and water resources and food production. We also conducted

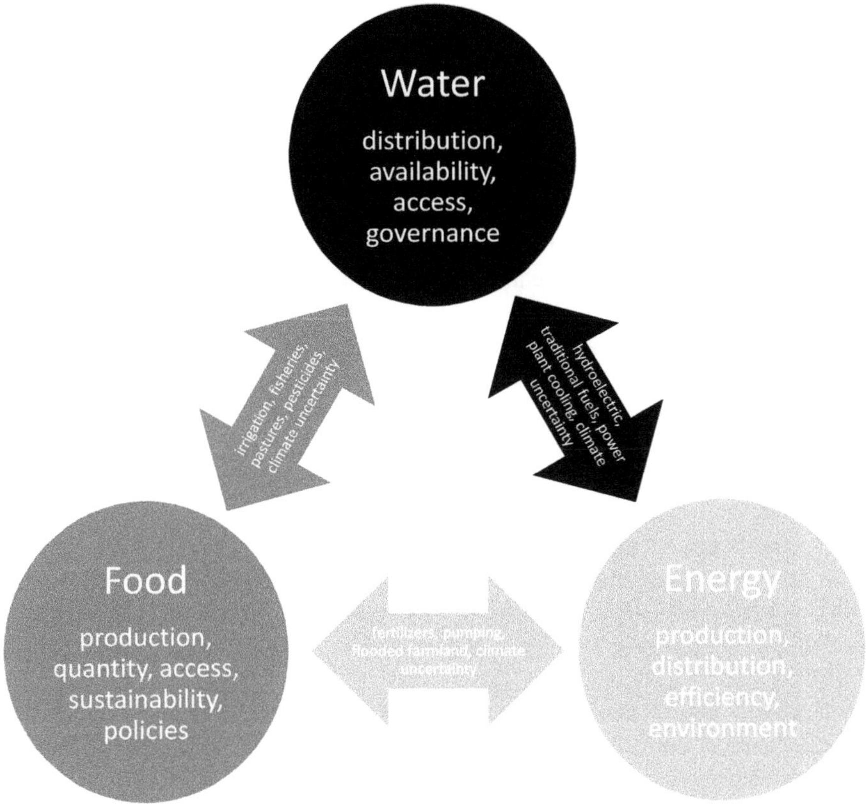

FIGURE 1.2 Integrative Approach to Natural Resource Security: Water-Energy-Food Nexus.[53]

case study analysis to elucidate the causal mechanisms and interactive dynamics. In this context, the book reflects the culmination of extensive fieldwork, including research visits and discussions with academics, government policymakers, and industrialists or other practitioners in the water, energy, and agricultural sectors in Azerbaijan, China, Georgia, Kazakhstan, Kyrgyzstan, the Russian Federation, Tajikistan, Turkey, and Uzbekistan.

The argument in the chapters that follow is that the legacies of the environmental and political misdeeds of Soviet policies and projects in its Eastern and Southern republics remain powerful today. Nonetheless, in much of Eurasia, there are signs of optimism and substantial reform which give rise to some evidence of the opportunity for broader prosperity and environmental sustainability through enhanced cooperation.

In Chapter 2, we trace the origins of European (especially German) dependence on Russia for energy resources. West Germany was seeking a final peace settlement to World War II, with the possibility of more stable relations with East Germany and the USSR. Germany and other European states also sought what they saw as potential for more stable alternative energy supplies after the OPEC oil boycott and price quadrupling in the wake of the Yom Kippur War. Russia sought a stable market for its oil and gas exports to earn hard currency, as well as pipeline construction technologies that the Germans offered. Was this the path to stabilizing Cold War tensions in Europe or imprudent dependence on a new unreliable supplier?

In Chapter 3, we carry this analysis forward to the second decade of the 21st century and assess the extent to which this European dependence on Russia for energy supplies has come to a close in the wake of the Russian invasion of Ukraine in February 2022. In some respects, it can be argued that the invasion has in fact been a game changer in European efforts to seek diversification of energy sources, including substantial commitment to renewables, a reconsideration of plans to reduce reliance on nuclear power generation, and increased import of liquified natural gas from the Middle East and the United States. There has been some "backsliding" in national plans to reduce consumption of domestically mined coal, including heavily polluting and carbon emitting soft coal, especially in Germany and Poland. This was probably inevitable given the dramatic curtailment of European imports of Russian oil and natural gas, by EU sponsored sanctions and Russian retaliation.

In Chapter 4, we explore the Water-Energy-Food Nexus and sustainable development efforts as they have evolved in the eight states of Central Eurasia: Armenia, Azerbaijan, Georgia, Kazakhstan, Kyrgyzstan, Tajikistan, Turkmenistan, and Uzbekistan. How independent are these countries economically, and how successful have they been in building broad-based prosperity and diversified and environmentally sustainable economies? Water scarcity is particularly dire in Kazakhstan, Turkmenistan, and Uzbekistan, given climate change and the legacy of disastrous Soviet agricultural policies, but there is evidence of agricultural diversification away from unsustainable cotton production and serious efforts to recover threatened inland fisheries and modernize aquaculture projects, particularly in Kazakhstan and Uzbekistan.

With Chapter 5, we begin to turn directly to the role of key external actors on the geopolitics and sustainable development of Central Eurasia. First, we ask how successful has the Russian Federation been in maintaining influence over its so-called "near abroad" successor states to the USSR, especially in Central Eurasia? Has the Russian invasion of Ukraine made a significant difference in the aspirations of key Eurasian states for efforts to secure long-term economic independence, sustainable development, and security?

In Chapter 6, we trace the significant rise of China as a key actor in Central Eurasia and beyond. In large part, this centers on the ambitious elements of Xi Jinping's Belt and Road Initiative (BRI), but it also derives somewhat from the foundation laid by earlier efforts of Kazakhstan and Turkmenistan to develop independent oil and gas trade with China, as well as the early development of the Shanghai Cooperation Organization (SCO) established in 2002. The chapter also explores the extent to which transborder water issues are growing in importance due to the rising Chinese agricultural development efforts in Xinjiang province which, for example, require substantial irrigation draw off from the Ili River before it reaches the Kazakh border. China-Kazakhstan negotiations have made little progress on equitable sharing of water after more than 15 years of meetings.

Chapter 7 concludes the examination of the influence of external actors on the geopolitics and sustainable development of Central Eurasia with a detailed discussion of Türkiye's pan-Turkism/neo-Ottoman ambitions on energy ties, particularly with the South Caucasus. The chapter also highlights the more prevalent role of security and energy concerns over ideational factors. Türkiye's transborder water relations, particularly to the South, are also discussed given the impact on Syrian and Iraqi agriculture and human consumption needs.

Chapter 8 concludes the volume with a summary analysis of the Water-Energy-Food nexus in relation to climate change and geopolitical impacts. We then assess the theoretical implications of our efforts to unravel the WEF concept and to broaden the security and geopolitical discussions of Central Eurasia.

Overall, this book helps to broaden our understanding of security and its intricate interactions with geopolitics and the environment. Military conflict is not neglected, but there is a clear interest in elucidating the role of energy and water resources and food security in shaping political, economic, and foreign policy dynamics. It offers an interdisciplinary approach by combining traditionally separated areas of academic research and offers a comparative perspective by linking distinct, but highly interrelated sub-regions and issues. The interaction of local, national, regional, and global security implications of scarce natural resource challenges are examined.

Notes

1 See, Joseph E. Stiglitz, *Globalization and Its Discontents Revisited: Anti-Globalization in the Era of Trump* (New York: W.W. Norton, 2018).

2 Central Eurasia for us (and many scholars of the region such as those commonly affiliating with the Central Eurasian Studies Association) is defined as the five successor states to the USSR in Central Asia (Kazakhstan, Kyrgyzstan, Tajikistan, Turkmenistan, and Uzbekistan) along with the South Caucasus successor states of Armenia, Azerbaijan, and the Republic of Georgia. Some scholars might add Afghanistan and perhaps territory within the Russian Federation or China (e.g., Xinjiang Province), but we shall focus largely on the Soviet successor states for the purposes of this analysis.

3 See, Jacob M. Landau, *Pan-Turkism: From Irredentism to Cooperation* (Bloomington: Indiana University Press, 1995); and M. Hakan Yavuz, *Nostalgia for the Empire: The Politics of Neo-Ottomanism* (Oxford: Oxford University Press, 2020).

4 Michael T. Klare, *Rising Power, Shrinking Planet: The New Geopolitics of Energy* (New York: Henry Holt, 2008).

5 See, International Energy Agency (IEA), "Energy Security: Reliable, Affordable Access to All Fuels and Energy Sources." Retrieved on February 2, 2022, from https://www.iea.org/topics/energysecurity.

6 IEA, "Energy Security."

7 European Commission, "European Green Deal Document 52019DC0640." Summary, May 10, 2020, p. 2.

8 The Security Council of Russian Federation, "The National Security Strategy of Russian Federation," May 12, 2009, Decree of the RF President No. 537, from http://mepoforum.sk/wp-content/uploads/2015/08/NDS-RF-2009-en.pdf

9 T. van de Graaf and J.D. Colgan, "Russian Gas Games or Well-Oiled Conflict? Energy Security and the 2014 Ukraine Crisis," *Energy Research & Social Science* 24 (2017): pp. 59–64; A.N. Stulberg, "Natural Gas and the Russia-Ukraine Crisis: Strategic Restraint and the Emerging Europe-Eurasia Gas Network," *Energy Research & Social Science* 24 (2017): pp. 71–85; Belgin San-Akca, S. Duygu Sever and Şuhnaz Yılmaz, "Does Natural Gas Fuel Civil War? Rethinking Energy Security, International Relations, and Fossil-Fuel Conflict," *Energy Research and Social Science* 70 (December 2020); Abibail Adcox, "Russia Doubles Profits from Oil and Gas Sales to EU during War." Citing Study by Centre for Research on Energy and Clean Air, April 28, 2022.

10 For some insightful studies regarding the impact of geopolitical factors in shaping energy politics in Eurasia, see Paul Kubicek, "Energy Politics and Geopolitical Competition in the Caspian Basin," *Journal of Eurasian Studies* 4, no.2 (2013): pp. 171–180; Emre Erşen and Mitat Çelikpala, "Turkey and the Changing Energy Geopolitics of Eurasia," *Energy Policy* 128 (2019): pp. 584–592; Pavel K. Baev and Indra Overland, "The South Stream versus Nabucco Pipeline Race: Geopolitical and Economic (Ir)Rationales and Political Stakes in Mega-Projects," *International Affairs* 86, no. 5 (2010): pp. 1075–1090; Per Högselius, *Red Gas: Russia and the Origins of European Energy Dependence* (Springer, 2012). See also: Valentine, Scott Victor, et al., *Empowering the Great Energy Transition: Policy for a Low-Carbon Future* (New York: Columbia University Press, 2019); and Dieter Helm, *Burn Out: The Endgame for Fossil Fuels* (New Haven, CT: Yale University Press, 2017); and Lester R. Brown, et al., *The Great Transition: Shifting from Fossil Fuels to Solar and Wind Energy* (New York: W.W. Norton, 2015).

11 See: Margarita M. Balmaceda, *Russian Energy Chains: The Remaking of Technopolitics from Siberia to Ukraine to the European Union* (New York: Columbia University Press, 2021).

12 Jason Bordoff, "Everything You Think about the Geopolitics of Climate Change Is Wrong," Op-ed., *Foreign Policy*, October 5, 2020. Available at: https://www.energypolicy.columbia.edu/research/op-ed/everything-you-think-about-geopolitics-climate-change-wrong.

13 See, for example, Miriam Lowi, *Water and Power: The Politics of a Scarce Resource in the Jordan River Basin* (New York: Cambridge University Press, 1993); Thomas Homer-Dixon, "On the Threshold: Environmental Changes as Cause of Acute Conflict," *International Security* 16, no. 2 (1991): 76–116; Nazli Choucri, *Global Accord: Environmental Challenges and International Responses* (Cambridge, MA: MIT Press, 1993).

14 Program on Water Governance. Water Security, 2012. Retrieved on April 22, 2019, from http://watergovernance.ca/projects/water-security/

15 Ibid.

16 Shlomi Dinar, "Water, Security, Conflict, and Cooperation," *SAIS Review* 22, no. 2 (Summer/Fall 2002): pp. 229–253; Patrick M. Morgan, *International Security: Problems and Solutions* (Washington, DC: CQ Press, 2006).

17 Barry Buzan, Ole Waever, and Jaap De Wilde, *Security: A New Framework for Analysis* (Lynne Rienner Publishers, 1998).

18 Bertrand Charrier and Paul Samson, *International Freshwater Conflict: Issues and Prevention Strategies* (Geneva: Green Cross International, 1997), p. 6; Thomas Homer-Dixon, "On the Threshold: Environmental Changes as Cause of Acute Conflict," *International Security* 16, no. 2 (1991): pp. 76–116; Astri Suhrke, "Environmental Change, Migration and Conflict: A Lethal Feedback Dynamic?," in *Managing Global Chaos: Sources of and Responses to International Conflict*, eds. C. Crocker, F. Hampson, and P. Aall (Washington, DC: United States Institute of Peace Press, 1996), p. 115; Harriet Critchley and Terry Terriff, "Environment and Security," in *Security Studies for the 1990s*, eds. R. Schultz, R. Godson, and T. Greenwood (Washington, DC: Brassey's, 1993): p. 328; Peter H. Gleick, "Water and Conflict: Fresh Water Resources and International Security," *International Security* 18, no. 1 (1993): pp. 23–41; 81–82, 92–93; Ian Rowlands, "The Security Challenges of Global Environmental Change," *The Washington Quarterly* 14, no. 1 (1991): pp. 99–114.

19 Raymond L. Bryant (ed.), *International Handbook of Political Ecology* (Northampton, MA: Edward Elgar Publishing, 2015); Peet, Robbins and Watts (eds.), *Global Political Ecology*.
20 Simon Dalby. *Anthropocene Geopolitics: Globalization, Security, Sustainability* (University of Ottawa Press, 2020). https://doi.org/10.2307/j.ctvx5w8dk.
21 Thomas Diez, Franziskus von Lucke, and Zehra Wellmann, *The Securitisation of Climate Change: Actors, Processes and Consequences* (Routledge, 2016).
22 See, George G. Eberling, *China's Bilateral Relations with Its Principal Oil Suppliers* (Lapham, MD: Lexington Books, 2017); Philip Andrews-Speed and Roland Dannreuther, *China, Oil and Global Politics* (London: Routledge, 2011); Anatole Boute, *Energy Security along the New Silk Road* (Cambridge: Cambridge University Press, 2021); Jim Krane, *Energy Kingdoms: Oil and Political Survival in the Persian Gulf* (New York: Columbia University Press, 2021); and Rajan Menon et al., *Russia, the Caucasus, and Central Asia: The 21st Century Security Environment* (Armonk, NY: Sharpe, 1999).
23 See, Paul J. Bolt and Sharyl N. Cross, *China, Russia and Twenty-First Century Global Geopolitics* (Oxford: Oxford University Press, 2018); Michal Lubina, *Russia and China: A Political Marriage of Convenience Stable and Successful* (Berlin: Verlag Barbara Budrich, 2017); and Ali Askerov, *Contemporary Russo-Turkish Relations: From Crisis to Cooperation* (Lapham, MD: Lexington Books, 2018).
24 Caroline Kuzemko, *The Energy Security and Climate Nexus: Institutional Change in the UK and Beyond* (London: Palgrave Macmillan, 2013).
25 T. Van de Graaf and J.D. Colgan, "Russian Gas Games or Well-Oiled Conflict? Energy Security and the 2014 Ukraine Crisis," *Energy Research & Social Science* 24 (2017): 59–64.
26 Daniel Yergin, *The New Map: Energy, Climate, and the Clash of Nations* (London: Penguin, 2020); and Daniel Yergin. *The Quest: Energy, Security, and the Remaking of the Modern World* (London: Penguin, 2011).
27 Agnia Grigas, *The New Geopolitics of Natural Gas* (Cambridge, MA: Harvard University Press, 2017).
28 Adam Stulberg, *Well-Oiled Diplomacy: Strategic Manipulation and Russia's Energy Statecraft in Eurasia* (Albany, NY: State University of New York Press, 2008).
29 See: Steve Levine, *The Oil and the Glory: The Pursuit of Empire and Fortune on the Caspian Sea* (New York: Random House, 2007); James Marriott and Mike Minio-Paluello, *The Oil Road: Journeys from the Caspian Sea to the City of London* (London: Verso, 2013).
30 See: Brahma Chellaney, *Water, Peace, and War: Confronting the Global Water Crisis* (Lanham, MD: Rowman & Littlefield Publishers, 2015); Walter Leal Filho and Dinesh Surroop, *The Nexus: Energy, Environment and Climate Change* (New York: Springer, 2017).
31 Piotr Dutkiewicz, et al., *Eurasia on Edge: Managing Complexity* (Lanham, MD: Lexington Books, 2018).
32 For the economic dynamics underlying energy politics with valuable historical insights, see: J. Hartshorn, *Oil Trade: Politics and Prospects*. Cambridge Energy and Environment Series (Cambridge, UK: Cambridge University Press, 1992).
33 United Nations Development Programme (UNDP), *Human Development Report 1994* (New York: Oxford University Press, 1994), pp. 23–25. Available at: https://hdr.undp.org/content/human-development-report-1994
34 Interviewed by Stephen Sackur on the BBC television program "Hard Talk," May 24, 2021. See also, Elizabeth Kolbert, *The Sixth Extinction: An Unnatural History* (New York: Henry Holt and Co., 2014).
35 G7 Climate and Environment Ministers Meeting. Available at: https://www.gov.uk/government/publications/g7-climate-and-environment-ministers-meeting-may-2021-communique
36 United Nations. *Sharm el-Sheik Implementation Plan.* Decision 1/CP.27. FCCC/CP/2022/10/Add.1. 17 March 2023.
37 Important elements of agricultural diversification include cattle production in the dryland regions and aquaculture development in several sub-regions. See: Jiquan Chen,

et al., "Sustainability Challenges for the Social-Environmental Systems across the Asian Drylands Belt," *Environmental Research Letters*, December, 2021.

38 See: S. Frederick Starr and Svante E. Cornell (eds.), *Uzbekistan's New Face* (Lanham, MD: Rowman and Littlefield, 2018).

39 Norman A. Graham, "The Prospect for Regional Governance of Inland Fisheries in Central Eurasia." Paper prepared for the FAO-MSU Global Conference on Inland Fisheries, January 2015, Rome. Published in select proceedings of the Conference in 2016, and "The Current Status and Future of Central Asia's Fish and Fisheries: Confronting a Wicked Problem" with Steven G. Pueppke and Talgat Uderbayev, published in special issue of *Water* on water management in Central Asia, August 2017.

40 See: Lei Zou, *The Political Economy of China's Belt and Road Initiative* (Singapore: World Scientific Publishing, 2018), especially Ch. 7 and the Conclusion.

41 Norman Graham, "The Prospect for Regional Governance of Inland Fisheries in Central Eurasia," 2015; The International Crisis Group, Asia Reports 34 *Central Asia: Water and Conflict* (2002); The *Curse of Cotton: Central Asia's Destructive Monoculture* (2005); World Bank, Europe and Central Asia Region. Water Energy Nexus in Central Asia: Improving Regional Cooperation in the Syr Darya Basin (Washington, 2004).

42 For a detailed inventory of European climate and energy policies, see: Leonardo Massai, *European Climate and Clean Energy Law and Policy* (London: Earthscan/Routledge, 2012). For specific earlier noncompliance challenges, see Chapter 19.

43 See: Norman A. Graham, Folke Lindahl, and Timur Kocaoglu, *Making Russia and Turkey Great Again? Putin and Erdogan in Search of Lost Empires and Autocratic Power* (Lanham, MD: Lexington Books/Rowman & Littlefield, 2021).

44 Kishore Mahbubani, *Has China Won? The Chinese Challenge to American Primacy* (New York: Public Affairs/Hachette Book Group, 2020).

45 Andrew Moravcsik, "Introduction: Integrating International and Domestic Theories of International Bargaining," in *Double-Edged Diplomacy: International Bargaining and Domestic Politics*, ed. Peter B. Evans, Harold Karan Jacobson, and Robert D. Putnam (Berkeley: University of California Press, 1993), pp. 3–41.

46 Robert Putnam, "Diplomacy and Domestic Politics: The Logic of Two-Level Games," *International Organization* 42, no. 3 (1988): pp. 427–460.

47 The U.S. Energy Information Administration, *International Energy Outlook 2019* (IEO2019). Available at https://www.eia.gov/todayinenergy/detail.php?id=41433.

48 Asit K. Biswas, *International Waters of the Middle East: From Euphrates-Tigris to Nile* (New York: Oxford University Press, 1994); Ali Çarkoglu and Mine Eder, "Domestic Concerns and the Water Conflict over the Euphrates-Tigris River Basin," *Middle Eastern Studies* 37, no. 1 (2001): pp. 41–71. doi:10.1080/714004364; John F. Kolars and William A. Mitchell, *The Euphrates River and the Southeast Anatolia Development Project* (Carbondale: SIU Press, 1991); Brahma Chellaney, *Water, Peace, and War: Confronting the Global Water Crisis* (Lanham, MD: Rowman & Littlefield, 2013).

49 Geoffrey D. Dabelko, "Environmental Security Heats Up," *Environ Change and Security Program* rep. 13 (2008): pp. vii–x. Available at, https://www.wilsoncenter.org/publication/foreword-environmental-security-heats; Jon Barnett, *The Meaning of Environmental Security: Ecological Politics and Policy in the New Security Era* (London: Zed Books, 2001); Michael Klare, *The Race for What's Left: The Global Scramble for the World's Last Resources* (New York: Metropolitan Books, 2012); Naomi Oreskes and Erik M. Conway, *The Collapse of Western Civilization: A View from the Future* (New York: Columbia University Press, 2014).

50 UNESCO, *Water Security*. Retrieved on January 13, 2017, from: en.unesco.org/themes/water-security.

See also: World Economic Forum, *Water Security: The Water-Food-Energy-Climate Nexus*, Intro and Ch. 1; and Steven G. Pueppke, Sabir T. Nurtazin, Norman A. Graham, and Jiaguo Qi, "Central Asia's Ili River Ecosystem as a Wicked Problem: Unraveling Complex Interrelationships at the Interface of Water, Energy, and Food," *Water*.

51 Douglas Broom and Keith Breene, "This Is Why Food Security Matters Now More Than Ever," *World Economic Forum*, November 23, 2020. Available at: https://www.weforum.org/agenda/2020/11/food-security-why-it-matters/
52 Christine Parthemore and Will Rogers, "Sustaining Security: How Natural Resources Influence National Security," *Center for a New American Security* (2010): p. 5.
53 An earlier version of this diagram was crafted by Professor Jiaguo Qi of Michigan State University who has been one of the key architects of WEF nexus studies globally, especially in Asia and Eurasia. See, for example, Pueppke, Nurtazin, Graham, and Qi, "Central Asia's Ili River Ecosystem in the Nexus of Water, Energy, and Food," *Water*, April 2018.

References

Andrews-Speed, Philip, and Roland Dannreuther. *China, Oil and Global Politics*. London: Routledge, 2011.

Askerov, Ali. *Contemporary Russo-Turkish Relations: From Crisis to Cooperation*. Lanham, MD: Lexington Books, 2018.

Baev, Pavel K., and Indra Overland. "The South Stream versus Nabucco Pipeline Race: Geopolitical and Economic (ir)rationales and Political Stakes in Mega-Projects." *International Affairs* 86, no. 5 (2010): 1075–1090.

Balmaceda, Margarita M. *Russian Energy Chains: The Remaking of Technopolitics from Siberia to Ukraine to the European Union*. New York: Columbia University Press, 2021.

Barnett, Jon. *The Meaning of Environmental Security: Ecological Politics and Policy in the New Security Era*. London: Zed Books, 2001.

BBC. (2021, May 27). "We Want to Inspire Advocates for Planet Earth." Retrieved on January 19, 2021, from https://www.bbc.co.uk/programmes/p09jkrf5

Biswas, Asit K. *International Waters of the Middle East: From Euphrates-Tigris to Nile*. New York: Oxford University Press, 1994.

Bolt, Paul J., and Sharyl N. Cross. *China, Russia and Twenty-First Century Global Geopolitics*. New York: Oxford University Press, 2018.

Bordoff, Jason. "Everything You Think about the Geopolitics of Climate Change Is Wrong." Op-ed., *Foreign Policy*. October 5, 2020. Available at: https://foreignpolicy.com/2020/10/05/climate-geopolitics-petrostates-russia-china/

Boute, A. *Energy Security along the New Silk Road: Energy Law and Geopolitics in Central Asia*. Cambridge Studies on Environment, Energy and Natural Resources Governance. Cambridge: Cambridge University Press, 2019.

Broom, Douglas, and Keith Breene. "This Is Why Food Security Matters Now More Than Ever." *World Economic Forum*. November, 23 2020. Retrieved on December 21, 2021, from https://www.weforum.org/agenda/2020/11/food-security-why-it-matters/

Brown, Lester R. *Full Planet, Empty Plates: The New Geopolitics of Food Security*. New York: W.W. Norton, 2012.

Brown, Lester R. et al. *The Great Transition: Shifting from Fossil Fuels to Solar and Wind Energy*. New York: W.W. Norton, 2015.

Bryant, Raymond L. (ed.), *International Handbook of Political Ecology*. Northampton, MA: Edward Elgar Publishing, 2015.

Buzan, Barry, Ole Waever, and Jaap De Wilde. *Security: A New Framework for Analysis*. Boulder, CO: Lynne Rienner Publishers, 1998.

Çarkoglu, A., and M. Eder. "Domestic Concerns and the Water Conflict over the Euphrates-Tigris River Basin." *Middle Eastern Studies* 37, no. 1 (2001): 41–71.

Charrier, Bertrand, and Paul Samson. *International Freshwater Conflict: Issues and Prevention Strategies*. Geneva: Green Cross International, 1997.

Chellaney, Brahma. *Water, Peace, and War: Confronting the Global Water Crisis*. Lanham, MD: Rowman & Littlefield, 2013.

Chen, Jiquan, et al. "Sustainability Challenges for the Social-Environmental Systems across the Asian Drylands Belt." *Environmental Research Letters*. December, 2021.

Choucri, Nazli. *Global Accord: Environmental Challenges and International Responses.* Cambridge: MIT Press, 1993.

Cooley, Alexander, and John Heathershaw. *Dictators Without Borders: Power and Money in Central Asia.* New Haven, CT: Yale University Press, 2017.

Corey, Johnson, et al. *Want, Waste or War?: The Global Resource Nexus and the Struggle for Land, Energy, Food, Water & Minerals.* New York: Routledge, 2015.

Critchley, Harriet, and Terry Terriff. "Environment and Security," in *Security Studies for the 1990s*, edited by R. Schultz, R. Godson, and T. Greenwood. Washington, DC: Brassey's, 1993.

Dabelko, Geoffrey D. "Environmental Security Heats Up." *Environmental Change and Security Program (ECSP) Report* 13, (2008).

Dalby, Simon. *Anthropocene Geopolitics: Globalization, Security, Sustainability*. Ottawa: University of Ottawa Press, 2020. https://doi.org/10.2307/j.ctvx5w8dk.

Dellecker, Adrian, and Thomas Gomart (eds.), *Russian Energy Security and Foreign Policy*. London: Routledge, 2011.

Diez, Thomas, and Franziskus von Lucke, Zehra Wellmann. *The Securitisation of Climate Change: Actors, Processes and Consequences*. London: Routledge, 2016.

Dinar, Shlomi. "Water, Security, Conflict, and Cooperation." *SAIS Review* 22, no. 2 (2002): 229–253.

Dinar, Shlomi (ed.), *Beyond Resource Wars: Scarcity, Environmental Degradation, and International Cooperation*. Cambridge, MA: MIT Press, 2012.

Dutkiewicz, Piotr, et al. *Eurasia on Edge: Managing Complexity.* Lanham, MD: Lexington Books, 2018.

Eberling, George G. *China's Bilateral Relations with Its Principal Oil Suppliers*. Lanham, MD: Lexington Books, 2017.

Erşen, Emre, and Mitat Çelikpala. "Turkey and the Changing Energy Geopolitics of Eurasia." *Energy Policy* 128 (2019): 584–592.

European Commission, European Green Deal. May 10, 2020. https://ec.europa.eu/info/strategy/priorities-2019-2024/european-green-deal_en

Filho, Walter Leal, and Dinesh Surroop. *The Nexus: Energy, Environment and Climate Change.* New York: Springer, 2018.

Gleick, Peter H. "Water and Conflict: Fresh Water Resources and International Security." *International Security* 18, no. 1 (1993): 79–112.

Graham, N. "The Prospect for Regional Governance of Inland fisheries in Central Eurasia," in *Freshwater, Fish and the Future: Proceedings of the Global Cross-Sectoral Conference*, 2015.

Graham, Norman A., Folke Lindahl, and Timur Kocaoglu. *Making Russia and Turkey Great Again? Putin and Erdogan in Search of Lost Empires and Autocratic Power.* Lanham, MD: Lexington Books/Rowman&Littlefield, 2021.

Graham, Norman A., Steven G. Pueppke, Talgat Uderbayev. "The Current Status and Future of Central Asia's Fish and Fisheries: Confronting a Wicked Problem." *Water* 9, no. 9 (2017): 701.

Grigas, Agnia. *The New Geopolitics of Natural Gas*. Cambridge: Harvard University Press, 2017.

Gustafson, Thane. *Wheel of Fortune: The Battle for Oil and Power in Russia.* Cambridge, MA: Harvard University Press, 2012.
Gutierrez, Eric. "Boiling Point: Issues and Problems in Water Security and Sanitation." *Water Aid International Site*, August, 1999. http://citeseerx.ist.psu.edu/viewdoc/download?doi=10.1.1.581.8862&rep=rep1&type=pdf
Hartshorn, J. E. *Oil Trade: Politics and Prospects.* Cambridge Energy and Environment Series. Cambridge: Cambridge University Press, 1993.
Hedlund, Stefan. *Putin's Energy Agenda: The Contradictions of Russia's Resource Wealth.* Boulder, CO: Lynne Rienner, 2014.
Helm, Dieter. *Burn Out: The Endgame for Fossil Fuels.* New Haven, CT: Yale University Press, 2017.
Högselius, Per. *Red Gas: Russia and the Origins of European Energy Dependence.* New York: Palgrave Macmillan, 2013.
Högselius, P. *Energy and Geopolitics.* London: Routledge, 2018.
Homer-Dixon, Thomas. "On the Threshold: Environmental Changes as Cause of Acute Conflict." *International Security* 16, no. 2 (1991): 76–116.
International Energy Agency (IEA). "Energy Security: Reliable, Affordable Access to All Fuels and Energy Sources." https://www.iea.org/topics/energysecurity
International Energy Agency. "World Energy Outlook 2006." October, 2020. https://iea.blob.core.windows.net/assets/390482d0-149a-48c0-959b-d5104ea308ca/weo2006.pdf
Jewell, Jessica. "The IEA Model of Short-Term Energy Security (MOSES) Primary Energy Sources and Secondary Fuels." *International Energy Agency.* (2011).
Jones, Alun, and Julian Clark. *The Modalities of European Union Governance: New Institutionalist Explanations of Agri-Environmental Policy.* Oxford: Oxford University Press, 2001.
Klare, Michael. *Rising Power, Shrinking Planet: The New Geopolitics of Energy.* New York: Henry Holt, 2008.
Klare, Michael. *The Race for What's Left: The Global Scramble for the World's Last Resources.* New York: Metropolitan Books/Henry Holt, 2012.
Klein, Naomi. *This Changes Everything: Capitalism vs the Climate.* New York: Simon and Schuster, 2014.
Kleveman, Lutz. *The New Great Game: Blood and Oil in Central Asia.* London: Atlantic Books, 2003.
Kolars, John F., and William A. Mitchell. *The Euphrates River and the Southeast Anatolia Development Project.* Carbondale, IL: SIU Press, 1991.
Kolbert, Elizabeth. *The Sixth Extinction: An Unnatural History.* New York: Henry Holt and Co., 2014.
Krane, Jim. *Energy Kingdoms: Oil and Political Survival in the Persian Gulf.* New York: Columbia University Press, 2021.
Kubicek, Paul. "Energy Politics and Geopolitical Competition in the Caspian Basin." *Journal of Eurasian Studies* 4, no. 2 (2013): 171–180.
Kuzemko, Caroline. *The Energy Security and Climate Nexus: Institutional Change in the UK and Beyond.* London: Palgrave MacMillan, 2013.
Landau, Jacob M., *Pan-Turkism: From Irredentism to Cooperation.* Bloomington: Indiana University Press, 1995.
LeVine, Steve. *The Oil and the Glory: The Pursuit of Empire and Fortune on the Caspian Sea.* New York: Random House, 2007.
Lowi, Miriam. *Water and Power: The Politics of a Scarce Resource in the Jordan River Basin.* New York: Cambridge University Press, 1993.
Lubina, Michal. *Russia and China: A Political Marriage of Convenience Stable and Successful.* Opladen: Verlag Barbara Budrich, 2017.

Mahbubani, Kishore. *Has China Won? The Chinese Challenge to American Primacy.* New York: Public Affairs/Hachette Book Group, 2020.
Manning, Robert A. *The Asian Energy Factor: Myths and Dilemmas of Energy, Security, and the Pacific Future*. New York: Palgrave, 2000.
Marriott, James, and Mika Minio-Paluello. *The Oil Road: Journeys from the Caspian Sea to the City of London.* London: Verso, 2012.
Massai, Leonardo. *European Climate and Clean Energy Law and Policy*. London: Earthscan/Routledge, 2012.
Mathews, Jessica. "Redefining Security." *Foreign Affairs* 68, no. 2 (1989): 162–177.
Memon, Rajan, et al. *Russia, the Caucasus, and Central Asia: The 21st Century Security Environment.* London: Sharpe, 1999.
Mirbabayev, Miryusif. *Concise History of Azerbaijani Oil.* Baku, AZ: SOCAR, 2008.
Moravcsik, Andrew. "Introduction: Integrating International and Domestic Theories of International Bargaining," in *Double-Edged Diplomacy*, edited by Peter B. Evans, Harold Karan Jacobson, and Robert D. Putnam. Berkeley: University of California Press, 1993.
Morgan, Patrick M. *International Security: Problems and Solutions*. Washington, DC: CQ Press, 2006.
Myers, Norman. "Environment and Security." *Foreign Policy* no. 74 (1989): 23–41.
Oreskes, Naomi, and Erik Conway. *The Collapse of Western Civilization: A View from the Future*. New York: Columbia University Press, 2014.
Parthemore, Christine, and Will Rogers. "Sustaining Security: How Natural Resources Influence National Security." *Center for a New American Security* (2010).
Porter, Gareth. "Post-Cold War Global Environment and Security." *The Fletcher Forum of World Affairs*, 14, no. 2 (1990): 332–333.
Program on Water Governance. *Water Security*. (2012). http://watergovernance.ca/projects/water-security/
Pueppke, Steven G., Sabir T. Nurtazin, Norman A. Graham, and Jiaguo Qi. "Central Asia's Ili River Ecosystem as a Wicked Problem: Unraveling Complex Interrelationships at the Interface of Water, Energy, and Food." *Water* 10, no. 5 (2018): 541.
Putnam, Robert. "Diplomacy and Domestic Politics: The Logic of Two-Level Games." *International Organization* 42, no. 3 (1988): 427–460.
Rowlands, Ian. "The Security Challenges of Global Environmental Change." *The Washington Quarterly* 14, no. 1 (1991): 99–114.
Russian Government. "National Security Strategy of the Russian Federation to 2020, Approved by Decree of the President of the Russian Federation 12 May 2009." http://mepoforum.sk/wp-content/uploads/2015/08/NDS-RF-2009-en.pdf
San-Akca, Belgin, S. Duygu Sever, and Şuhnaz Yılmaz. "Does Natural Gas Fuel Civil War? Rethinking Energy Security, International Relations, and Fossil-Fuel Conflict." *Energy Research and Social Science* 70 (December 2020).
Shultz, Richard H., Roy Godson, and Ted Greenwood. "Environment and Security," in *Security Studies for the 1990s*. Washington, DC: Brassey's, 1993.
Starr, S. Frederick, and Svante E. Cornell (eds.), *Uzbekistan's New Face*. Lanham, MD: Rowman and Littlefield, 2018.
Stiglitz, Joseph E. *Globalization and Its Discontents Revisited: Anti-Globalization in the Era of Trump*. New York: W.W. Norton, 2018.
Stulberg, A.N. *Well-Oiled Diplomacy: Strategic Manipulation and Russia's Energy Statecraft in Eurasia*. New York: SUNY Series in Global Politics, 2007.
Stulberg, A.N. "Natural Gas and the Russia-Ukraine Crisis: Strategic Restraint and the Emerging Europe-Eurasia Gas Network." *Energy Research & Social Science* 24 (2017): 71–85.

Suhrke, Astri. "Environmental Change, Migration and Conflict: A Lethal Feedback Dynamic?," in *Managing Global Chaos: Sources of and Responses to International Conflict*, edited by C. Crocker, F. Hampson, and P. Aall. Washington, DC: United States Institute of Peace Press, 1996.

The International Crisis Group Asia Reports 34. "Central Asia: Water and Conflict." May 30, 2002. https://www.crisisgroup.org/europe-central-asia/central-asia/uzbekistan/central-asia-water-and-conflict

The International Crisis Group Asia Reports 93. "The Curse of Cotton: Central Asia's Destructive Monoculture." February 28, 2005. https://www.crisisgroup.org/europe-central-asia/central-asia/tajikistan/curse-cotton-central-asias-destructive-monoculture

Tunsjo, Oystein. *Security and Profit in China's Energy Policy: Hedging Against Risk.* New York: Columbia University Press, 2013.

UK Government. "G7 Climate and Environment Ministers Meeting." May 21, 2021. https://www.gov.uk/government/publications/g7-climate-and-environment-ministers-meeting-may-2021-communique

United Nations. *Sharm el-Sheik Implementation Plan.* Decision 1/CP.27. FCCC/CP/2022/10/Add.1. 17 March 2023.

United Nations Development Programme (UNDP). *Human Development Report 1994*. New York: Oxford University Press, 1994. https://hdr.undp.org/content/human-development-report-1994

U.S. Energy Information Administration. "International Energy Outlook 2017 (IEO2017)." September 14, 2017. https://www.eia.gov/todayinenergy/detail.php?id=32912

U.S. Energy Information Administration. "International Energy Outlook 2019 (IEO2019)." September 24, 2019. Retrieved on December 15, 2021, from https://www.eia.gov/todayinenergy/detail.php?id=41433

Valentine, Scott Victor, et al. *Empowering the Great Energy Transition: Policy for a Low-Carbon Future*. New York: Columbia University Press, 2019.

van de Graaf, T., and J. D. Colgan. "Russian Gas Games or Well-Oiled Conflict? Energy Security and the 2014 Ukraine Crisis." *Energy Research & Social Science* 24 (2017): 59–64.

World Bank. "Water Energy Nexus in Central Asia: Improving Regional Cooperation in the Syr Darya Basin (English)." http://documents.worldbank.org/curated/en/122241468232522184/Water-energy-nexus-in-Central-Asia-improving-regional-cooperation-in-the-Syr-Darya-Basin

World Economic Forum. "Water Security: The Water-Food-Energy-Climate Nexus." 2011. http://www3.weforum.org/docs/WEF_WI_WaterSecurity_WaterFoodEnergyClimateNexus_2011.pdf

Yavuz, M. Hakan. *Nostalgia for the Empire: The Politics of Neo-Ottomanism.* Oxford: Oxford University Press, 2020.

Yergin, Daniel. *The Quest: Energy, Security, and the Remaking of the Modern World.* London: Penguin, 2011.

Yergin, Daniel. *The New Map: Energy, Climate, and the Clash of Nations*. London: Penguin, 2020.

Zou, Lei. *The Political Economy of China's Belt and Road Initiative.* Singapore: World Scientific Publishing, 2018.

2

THE ENERGY SECURITY OF EUROPE

How Did Europe Become So Dependent on Russia?

The development of a strategy for energy security in Europe has a long and tortured history. Tortured in part because it is tied to the larger context of East-West relations in Europe and Europe's (especially Germany's) efforts to promote and maintain economic strength while trying to have a normal relationship with the Russian Federation. Energy-rich states in Central Eurasia (especially Azerbaijan, Kazakhstan, and Turkmenistan) became important to this relationship in the last years of the 20th century, as Europe sought to diversify its energy sources. But the dependence on Russian supply remained substantial well beyond 2010.

The events in Ukraine since 2014 along with the Russian meddling in Western elections and politics complicate these challenges, but the roots of energy dependence go back to Willy Brandt's Ostpolitik in pursuit of a treaty framework for a peace settlement after World War II and eventual German reunification. Not surprisingly, domestic politics and economic interests were important to the process and outcomes of the negotiations. One can argue that the challenge of reducing dependence on Russian oil and gas supplies also has its roots in the 1970s, if not earlier. Russia was eager to sell gas to Europe then, given its need for hard currency and the ability to purchase Western technology, and Europe was ready to diversify its energy import sources away from the Middle East. Prioritizing domestic economic interests over security and foreign policy principles was evident early on and was a key factor in Europe's reluctance to limit economic ties with Russia, even when it behaved badly (as an unreliable supplier—at minimum—or as a serious security threat).

Trade with the Soviet Union was both an opportunity for German industry and a means of encouraging normalization and support for the post-World War II settlement. With the steep rise in energy prices and embargo stemming from OPEC's response to the Yom Kippur Arab-Israeli war in 1973, and general questions about the stability of that supply source, Europe was anxious to reduce its over-dependence

DOI: 10.4324/9781003373735-2

on the Middle East. The Soviets were anxious to market natural gas to Europe, so a meeting of the minds occurred readily, despite the severe warnings to the Germans about "exchanging dependencies" coming from Washington. These challenges remain important today, although the question of diversification is now (actually since at least 2006) focused on reducing dependence on Russian sources of gas and oil. This is not easy for Europe to manage, but it has become both more necessary and more feasible recently, with the Russian invasion of Ukraine and its increasing pressure on other successor states to the Soviet Union, as well as changes in the global energy market. Unwillingness to move away from what James Sheehan termed Europe's preoccupation with becoming (only) a civilian super-state, have held it back, along with its tendency to put short-term commercial interests before medium to long-term security interests.[1]

In Search of a Peace Treaty and Energy Security

Between the German surrender on May 8 and the Japanese surrender on September 2, 1945, the Allies met at Potsdam to continue discussions of issues, the resolution of which were essential to a settlement. Because of disagreements between the Soviet Union and the United States, the only major decision reached with respect to Germany was to place three areas under Polish administration: Danzig, the southern portion of East Prussia, and the German territory lying east of the Oder and Neisse rivers. For the Western Powers, the Polish administration was regarded as provisional, pending a final settlement with Germany; but Poland and the Soviet Union regarded it as permanent. The Potsdam powers also decided that Germany was to be subject to a centralized occupation authority and treated as a single economic unit, but this decision was soon abandoned.

A variety of Western and Soviet proposals and counterproposals for Germany were made in the post-war period, but they were usually mutually exclusive. The American proposal for a four-power treaty in 1946 to guarantee the demilitarization of Germany was rejected by the Soviets, who proposed four-power control of the Ruhr, a proposal quickly rejected by the West. Disagreement over the status of Berlin became particularly bitter, erupting in two tense crises in 1948 and 1958. A conference on reunification was held in Berlin from January 25 to February 18, 1954. The negotiations failed, however, and serious talks concerning Germany did not resume for five more years.

In the second Berlin crisis (1958–1962), during which the Wall was erected, the Soviets threatened to negotiate a separate peace treaty with East Germany and terminate Western access to Berlin. Another Foreign Ministers Conference was held in Geneva in 1959. Like the Berlin Conference of 1954, it failed. Little progress toward settlement came until the late 1960s when domestic political considerations made reaching a settlement more attractive to West Germany and the Soviet Union.

Independent West German moves to relax tensions were not made until the Erhard and Kiesinger administrations in the years 1963–1969. Individual Eastern

European countries were the main targets in these efforts. However, only when Willy Brandt succeeded Kiesinger as Chancellor was there real success in resolving some of the difficult issues. Brandt's Chancellorship was based on a parliamentary coalition between the Social Democratic Party (SPD) and the Free Democratic Party (FDP), which came to power as a result of the September 1969 Bundestag elections. During the campaign, the SPD called for negotiations with the Soviet Union leading to an agreement that renounced the use of force. A few weeks before the election, Kiesinger revealed that the Soviets had offered to negotiate such a pact, an offer apparently aimed at helping the SPD campaign. On October 28, the occasion of Brandt's first policy statement, the new Chancellor told the Bundestag that he was willing to seek an arrangement with the German Democratic Republic (GDR) short of full diplomatic recognition. According to his proposal, the Federal Republic would formally agree to respect the GDR's territorial integrity as a separate state within the German nation. This new concession elicited a positive response from the East German leadership, and by the end of May 1970, the Chancellor had met twice with Prime Minister Willi Stoph for discussions concerning Brandt's demarche. Indeed, during that month, four separate yet interconnected sets of negotiations, dealing with Berlin, Poland, the two Germanies, and the Soviet Union, were going on concurrently.[2] The fall of 1971 had been marked by the emergence of CPSU First Secretary Leonid Brezhnev as the key Soviet official in efforts to achieve European settlements, as he assumed overall responsibility for the Soviet Union's relations with the United States and Western Europe. The Soviets pressed East Germany to reach agreement with the West as soon as possible.

Domestic politics in the Federal Republic of Germany (FRG) became significant as Brandt sought to have the Soviet and Polish friendship treaties ratified by the Bundestag. Despite defections from his majority and considerable pressure from the CDU, which wanted to secure more concessions from the GDR, the treaties were finally ratified on May 17, 1972 by a bare majority. Unanimous ratification by the Presidium of the Supreme Soviet quickly followed. On April 7, the FRG and the USSR, continuing their progress in economic cooperation, initialed a long-term trade agreement.

With the West German ratification of the Friendship Treaty, a Soviet precondition had been satisfied. The final Berlin settlement was signed by the four powers on June 3. West Germany and the GDR then began negotiations toward the establishment of formal relations. Talks were completed, and an agreement was made public on November 8, 1972, a little more than a week before the new Bundestag elections, which resulted in a considerably larger majority for the SPD-FDP coalition. The two Germanies signed their treaty on diplomatic relations on December 21, 1972. It was ratified by the Bundestag on June 6, 1973.

With little promise for significant additional gains from Ostpolitik in 1974, Brandt's efforts turned more toward domestic reform. The possibility of a settlement leading to an eventual reunification of Germany had been all but forgotten.

Unfortunately, by the time of Brandt's premature resignation in May 1974, only limited progress had been made in domestic reform.

There is little doubt that the most important factors for change in West Germany's policies with respect to reaching a settlement with the Soviet Union and East Germany derived from the domestic political situation in the Federal Republic. Indeed, it can be argued that domestic political considerations predominated over foreign policy or security considerations in the minds of West German policymakers.[3] As noted earlier, the key change (resulting from the September 1969 elections) was the formation of the SPD-FDP coalition that brought Willy Brandt to the post of Chancellor. The SPD campaign emphasized the need for reducing tensions in Europe, an emphasis duly recognized by the Soviet leadership, as the following exchange of telegrams between Soviet Premier Aleksey Kosygin and Brandt reveals. Kosygin wrote

> Accept my congratulations and best wishes on the occasion of your election as head of the government of the Federal Republic of Germany.
>
> Allow me to express the hope that the activity of your government will mark an improvement in the relations between the FRG and the Soviet Union. Such an improvement and good cooperation between our states would unquestionably be of great importance for the relaxation of tension in Europe and in the interests of a peaceful future for both the Federal Republic and the other European countries.

On October 24, Brandt thanked Kosygin and wrote:

> I agree with you that an improvement in the relations between our countries will be of great importance for a lasting peace in Europe. If both sides show the will to achieve positive results and commence this process, I shall be filled with a feeling of confidence.[4]

Secretary Brezhnev had expressed similar hopes in a speech in the GDR earlier that month. It was soon apparent that Brandt was willing to go considerably further than any previous Chancellor, in meeting the terms of the Soviets and East Germans, interpreting the slight margin of his victory as a mandate for attempting conciliation.

It seems reasonable to conclude that shifts in public opinion and partisan political maneuverings made it possible for Brandt and Egon Bahr, his chief negotiator for Ostpolitik, to do what they wanted, and indeed what they had planned to do in the first place. However, idiosyncratic and individual influences were of some importance. For example, Graham Hovey argued that Brandt's Eastern objectives followed logically from two aspects of his background. First, he desired to demonstrate that he was really German and was concerned about Germany's international position. Brandt had been absent from Germany during the whole World War II period, and he seems to have been quite sensitive to the critical remarks referring to his absence. The second aspect was Brandt's traumatic experience as Mayor of West Berlin during the erection of the Wall in 1961.[5] These two factors led Brandt to the conclusion that the Federal Republic could not rely on other states to find a solution to the German problem. Brandt's continued interest in reducing tensions in Europe was evident in his speeches and reports to the nation, which demonstrated flexibility and definite sense of direction.[6]

West German economic needs served as a domestic factor affecting Ostpolitik in three ways. First, it is clear that West Germany was dependent on foreign trade to a higher degree than most countries; and according to an analysis by the Institute of German Industry, this dependence was increasing. Uncertainty about the continued adequacy and availability of Western markets, coupled with this dependence led to some business pressure for an opening of trade with the East.[7] Negotiation of the Moscow-Bonn Friendship Treaty was required in order to establish a more auspicious trading climate, as Otto Von Amerongen, president of the West German Chamber of Industry and Commerce, noted in August 1970.[8] A specific example of this relationship was the West German interest in obtaining a long-term trade agreement to secure natural gas. The successful completion of this deal was closely connected with the conclusion of the Moscow-Bonn treaty.[9] The Soviets argued that the agreement was in keeping with their desire to negotiate economic pacts based on mutual advantage. They noted that the slumping West German steel industry had been stimulated by contracts for the supply of much needed large-diameter pipe to the USSR.[10] Indeed, trade supported by West German economic prowess in key industrial and technological areas served to stimulate the success of Ostpolitik and provided Bonn with a real bargaining tool.[11] On the negative side, the sectors of the FRG economy that had not performed well (e.g. the steel industry, some consumer goods sectors, and some agricultural sectors), and problems of recession and inflation, induced the coalition leadership to attempt to gain foreign policy successes in order to placate and distract the electorate.[12] However, economic problems forced the government, even in the face of domestic opposition, to adopt unpopular measures, including comparatively heavy taxation.[13]

There were significant external security motivations for the Soviets to move toward a peace settlement in Central Europe. However, there were also internal (largely economic and technological) factors that led to the coincidence between the Soviet relaxation toward the West and Brandt's Ostpolitik. By the late 1960s, the Soviet Union had reached strategic parity with the United States. Writing in 1972, C. G. Jacobsen stressed: "[The] Soviet Union has achieved full security vis-à-vis any exterior military threat."[14] Jacobsen's point may be a bit overdrawn, but he went on to add that the real question was whether security against internal threats had also been achieved. The answer seemed to be in the negative. Economic stagnation and technological inadequacy together with rising consumer demands led the Soviets to undertake improvement of relations with the West to derive benefits from economic and technological cooperation. The opening to the West in turn threatened to induce internal instability because it could expose the Soviet citizenry to new (and subversive) ideas. As Wolfgang Leonhard pointed out, internal repression in order to prevent subversion was one consequence of Soviet détente policy.[15] Moreover, concern for political instability was accentuated by the rise of dissension within the Politburo and Central Committee over the new economic and foreign policies, a development that tended to temper détente.

Strategic parity, however, did make the Soviets more confident in dealing with their foreign "opponents," especially the lesser ones like West Germany. Moreover, it led to changes in priorities that had the effect of giving domestic needs more weight in Soviet decision-making. Demands by the Soviet citizenry for consumer goods, for example, were given greater consideration after 1968. The overdue ninth Five-Year Plan, released in February of 1971, but not approved by the Politburo until October, placed significant stress on the production of consumer goods. Delays in drafting and approval were, in part, due to the objections of some Soviet leaders, especially Brezhnev, that early drafts of the plan had not placed enough stress on consumer goods.[16] Divisions in the leadership over other questions of policy and deficiencies in the performance of the Soviet economy were additional reasons. There is some evidence to suggest that economic problems resulted from the strain of the rapid buildup of strategic weapons in the Sixties that had produced Soviet parity with the United States. In any event, the problems were openly recognized by the Soviet leadership and severely criticized in public by Brezhnev.[17]

Demands for consumer goods, shortages of hard currency, and problems resulting from a lack of technological innovation could not be met by merely expanding trade with COMECON members, which was, in any case, a further drain on the capacities of the Soviet economy. Thus, trade with the more technologically advanced West had to be expanded; and this in turn required a policy of détente, which had been justified, both in the Soviet press and in resolutions of the Soviet Council of Ministers and the Central Committee of the Communist Party of the Soviet Union (CPSU), as means needed for improving the Soviet economy and benefiting the Soviet consumer.[18]

As Brandt noted in December 1970, one of the main reasons for the change in Moscow's German policy was desire for broadening economic and technological contacts with the more advanced Western industrial countries.[19] And indeed, trade and economic cooperation grew to the point where West Germany became the USSR's largest Western trading partner. The Soviets were particularly interested in obtaining steel comparable in quality to the large-diameter steel pipe secured on a long-term basis in the 1970 natural gas deal. The Soviet Union had become the world's largest producer of iron, but production lags and technological limitations limited the production of high-quality metal. West Germany entered into an agreement to construct a large Soviet metallurgical combine to help solve this problem.[20]

East Germany, in contrast, did not seem to need increased trade as much as the USSR and other Eastern European Countries. The 1960s were years of significant economic growth for the GDR—achieved without reliance on Western technology, capital, markets or goods, and without moderating its non-conciliatory stand on East-West relations. In addition, East Germany's total trade picture was considerably better than her neighbors to the East. It led them in exports and imports, and it maintained a favorable balance of trade since 1950.[21] Finally, East Germany enjoyed tariff-free access to the markets of the European Economic Community because of the prevailing convention that the GDR is not a country separate from the FRG.[22] (East Germany's demand for full diplomatic recognition by the FRG would have jeopardized this status, if granted.)

The West German economy benefited to some extent by increased trade and investment in the Soviet Union and Eastern Europe, and Soviet natural gas deals and joint ventures for the construction of atomic power stations helped the FRG diversify its domestic energy sources. The first impact of the settlement on the Soviet domestic situation, which also varied over time, was on the economy. Initially, the progress in economic cooperation, trade, and investment seemed quite satisfactory to the Soviets. A wide variety of economic, scientific, and technological agreements followed upon the signing of the Moscow-Bonn Friendship Treaty. In addition, there was a noticeable increase in the quantity of consumer goods imported from the West. In 1973, Soviet trade jumped 16%.[23] However, the Soviets expressed disappointment about the economic benefits they gained, particularly the relatively meager level of West German investment in the USSR. The Soviets had been seeking large amounts of capital at very low interest rates. West German reluctance persisted despite the fact that the Soviets discreetly made it clear on different occasions that they would trade political concessions and compromises on Berlin for more favorable economic agreements.[24]

The second important consequence of the settlement, and indeed of the whole Soviet policy of détente, was a tightening in internal security. Censorship and control of artists, writers, and dissenters increased after the adoption of Brezhnev's détente policies. This repression can be viewed as a concession to elements in the Soviet government, such as the secret police and military, which articulated their concern over the possible detrimental effects of détente.[25]

What is clear now, but was not so clear during Brezhnev's rule (at least to "outsiders"), is that the Soviet Union was heading toward economic crisis by the early 1980s, at the latest. A complex combination of the internal contradictions and deficiencies of the Soviet centrally planned command economy and the competition from rapid technological change and economic growth in many other parts of the globe proved too much for the Soviets to manage. Gorbachev's *Perestroika* and *Glasnost* reforms were well-intentioned—clearly the dismal state of the Soviet economy had gotten Gorbachev's attention, but these "reforms" were not enough and indeed may have hastened the collapse in some respects.

Per Högselius has provided perhaps the most detailed assessment of the origins of European energy dependence on Russian natural gas, especially beyond the

German case. He described it in terms of four distinct phases: (1) the late 1950s when it was uncertain that Europe would be a good market since in contrast to North America, natural gas was not used widely given the lack of domestic reserves. By 1964, though, the USSR had made its decision to become a gas exporter, and the key markets chosen were Italy, France, Austria, Finland, and Sweden; (2) after the 1968 Soviet invasion of Czechoslovakia, there was concern the West would be reluctant to import from the USSR, just as the first Austrian imports were about to start, but Europeans seemed to believe in fact that the trade would lead to better Soviet behavior; (3) the 1973/1974 Arab oil embargo and price raise stimulated new European customers and expansion of orders from the original ones; and (4) the chaos of the collapse of the Soviet Union and state socialism in Eastern Europe raised alarms about the growing dependence, but instead of cutting imports, the Europeans sought to manage the chaos as much as possible, including the challenge of newly independent transit countries—Belarus and Ukraine.[26]

A crucial feature in the Högselius argument is the extent to which Europeans found ways to justify the growing dependence and play down the potential security implications. One can argue that much of this was reasonable in the sense that these economic ties might have a positive impact on reforming the USSR and reducing Cold War tensions. But it is hard not to see the overwhelming business interests at work, trumping possible security concerns. The fact that the trade deals often involved "countertrade" in that the import bill was compensated for by lucrative large-diameter steel pipe and compressor equipment exports. Even the 1979 Soviet invasion of Afghanistan did not dissuade the Europeans from completing the massive Yamal pipeline deal; geopolitics at this stage would not get in the way of business and "secure" energy supplies, even in the face of substantial opposition from the US led by President Reagan.[27]

There were some efforts to reduce the prospect for the Soviets to use gas dependence as a weapon, or at least to manage the prospect for targeted disruptions by diversifying the pipelines into a complex network. The common view that seemed to matter most was that it was not in the interests of the USSR or indeed the Russian Federation to threaten this lucrative source of hard currency.[28] Moreover, there was little concern, at least up to the point of the book's publication in 2013 that military security issues were determinative. Högselius concluded:

> The attempts from the side of the United States to prevent Western Europe from cooperating with the communist bloc became a major hallmark of the Cold War. During most of the Cold War period, Washington preferred a divided Europe and sought, instead, to favor a tightly integrated mini-Europe in the West, with strong links to North America. West European countries themselves were less inclined to give up their traditional Eastern relations for the sake of ideological and military considerations. To judge from the material presented in this book, most Europeans regarded a much more open Europe, with large-scale flows of

> energy and technology between East and West, as the natural and historically justified path.[29]

The Soviet collapse in 1991, interestingly, had the remarkable impact of stimulating stronger economic ties between Europe and the successor states to the USSR, especially the Russian Federation. At the same time, the rise of Vladimir Putin would ultimately lead to a serious re-think on energy security.

Dependence on Energy Exports and the Collapse of the USSR

There is no space in this volume for a thorough treatment of the political and economic elements of the collapse of the USSR. Fortunately, there is no shortage of serious treatments of the complexities involved, from a variety of perspectives and disciplines.[30] On the one hand, the actual collapse (and especially the depth of economic weakness underpinning it) came as a shock to most analysts—even those privy to classified intelligence. It is not easy to build a consensus on the range of factors involved, particularly on their relative weight as determinants, but it is reasonable to summarize some common themes that seem important to most assessments. First, there were the evident characteristics of the centrally planned, command economy. Poorly informed centralized decisions were made for millions of minute product prices by GOSPLAN in Moscow (studiously avoiding insights that "market demands" might provide). This system worked well during the remarkable Soviet efforts to marshal resources in this command economy for high priority projects requiring substantial manpower and construction—e.g., heavy industrialization effort, armament production for World War II, large-scale canal digging, damming the Volga, the Soviet space program, large-scale ICBM production. But there was little scope for individual drive and innovation—little or no reward, indeed disincentives at times in normal, consumer product manufacturing. Consequently, an enormous and growing technological gap opened up with the West, clear to all by the 1980s, especially with the onset of the microelectronics revolution. Incentives were in the wrong direction, emphasizing, for example, production quantities—meeting the plan—over quality and variety; Grigory Yavlinsky argued, after observing the Soviet mining industry, that production quantities were also poor. The workers were not working, because there was no incentive to work hard.[31]

The Soviet system demonstrated a striking inability to manufacture competitive consumer products for export or even satisfactory domestic consumption. The authoritarian Soviet system—with nearly complete state ownership of property—left no room for entrepreneurial innovation and competitive commercial drive. There of course, contrary to Marx, was no evidence of the state "withering away" in this respect, indeed there was no real plan for reducing the role of the oppressive state machinery. There was a continuing focus instead on raw material exports (esp. oil and gas) to earn hard currency for necessary imports. This made the energy supply relationship evolving with Germany and Western Europe a high priority.

Economic policy making, indeed governance generally, lacked transparency and legitimacy; there was little or no real popular input on priorities. Secretary Brezhnev's massive spending priority to strategic defense to catch up with the United States largely went unchallenged; indeed, the true extent of the expenditure was an intriguing mystery even to Western intelligence agencies well into the 1980s. Robert Strayer, for example, placed much emphasis on Brezhnev's drive to catch up with the United States after the embarrassment of the Cuban Missile Crisis, as a causal factor in the Soviet collapse.[32] Others, while emphasizing that spending on strategic nuclear weaponry did essentially bankrupt the Soviet economy, noted that the nail in the coffin was the Soviet willingness to believe that President Reagan's Strategic Defense Initiative in 1983 might actually make much of Brezhnev's spending for naught—leading to a whole new round of arms race competition (and expenditure) potentially.

Soviet dependence on oil and gas income—with the corollary failure to diversify the industrial base—was key. Soviet and sympathetic conspiracy theorists drew attention to this dependence with recurring reports that the United States and Saudi Arabia had colluded to push oil prices down, thus depressing Soviet income. The effort reportedly was to punish the Soviets for the 1979 invasion of Afghanistan; the facts of the situation are hard to unravel, but it is clear that oil prices were low through much of the 1980s. The lack of export income in return made it hard for the Soviets to purchase needed grain from the West; the Soviet collective farm agricultural system continued to fail to perform adequately.[33] The US-Saudi collusion argument resurfaced in 2014 as a reported strategy to punish Russia for the annexation of Crimea. The fact that serious reform was needed seems not to have been in dispute to most analysts.[34]

Mikhail Gorbachev clearly recognized that serious reforms were needed. It is possible that a different approach might have saved the Soviet Union in some measure. But his chosen reform efforts were too little and too late and unleashed (destructive) forces that could not be countered. Gorbachev's reforms included an attack on Party (CPSU) procedures and elite personnel (the nomenklatura) without condemning communist ideology and the associated utopian dream; he encountered resistance in Party hierarchy, which maintained a strong interest in the privileges they held in the status quo.

He called for *Glasnost* and *Perestroika* (openness and restructuring); the former helped to "let the cat out of the bag" creating new awareness of what was wrong with the system and allowing serious opposition to blossom and the rise of alternative perspectives instead of unified national focus on substantial restructuring; it legitimized dissent while adding to delegitimization of the state socialist system. Aside from marginalizing some aspects of the role of the CPSU, key elements were some decentralization of economic planning and decision-making and the Law on Cooperatives, which permitted some private ownership and entrepreneurial activity.[35] Beyond these modest reforms, Gorbachev sought to reduce the pressure on the economy through arms control agreements with the United States. President Reagan's Strategic Defense Initiative and his interest in following through on the nuclear weapons modernization efforts begun by President Carter had raised the importance of curtailing the expensive arms race that Brezhnev had helped to fuel.

The reform effort by Gorbachev did unambiguously fail to revive the Soviet economy. Committed British Socialist Tony Benn famously argued that it was now possible to actually get Communism right! The August 1991 coup attempt which in effect brought down Gorbachev while repudiating bungled efforts by the hardline members of the CPSU to turn back the clock, also removed Benn's dream from contention. The attempt lasted only two days, but the Gorbachev reforms, including a negotiated new union treaty that decentralized power to the republics, were discredited—too little, too late. The Soviet Union voluntarily disintegrated, and the 15 successor states, including Russia embarked on a tortuous dual transition of political and economic transformation. Boris Yeltsin was unable to manage this effectively, setting the conditions that made it possible for Vladimir Putin to begin a gradual transformation which included regaining state control of natural resource assets, as well as progressive centralization of his political power. But the legacy of industrial weakness, poverty, and inequality in Russia has been an issue ever since. Remarkably, Putin's popularity largely has remained substantial nonetheless, while potential opponents were effectively constrained.[36]

From the Emergence of Russia as an Energy "Juggernaut" to a New Form of Economic Dysfunction

The depth of the economic decline and social dislocation to which the Russian Federation fell was hard to believe for most not experiencing it. On the surface, Russia appeared to have everything required to prosper under a new liberal economic order: a well-educated population, enormous reserves of natural resources, particularly oil and gas, an extensive transportation network with related infrastructure in place, and the prospect for access to markets well beyond its former key trade partners in Eastern Europe. Russia finally acceded to formal WTO membership toward the end of 2012, after considerable ambivalence in its political leadership—and 18 years of negotiation. But the pervasive corruption that was evident in the early years of transition has only deepened and remains the most startling characteristic of a political economy, with concomitant weakness in the rule of law that still fails to meet the basic needs of large segments of the population. The World Bank reported in 1995 and 2005 substantial poverty and weakness in health and social services. Life expectancy declined significantly after the collapse of the Soviet Union, while income inequality grew dramatically. There existed a very wealthy elite, perhaps only 0.3–1.5% of the population.[37]

Initial progress to address these challenges fell victim to the 1998 Russian financial crisis and subsequent policy to decrease social and health service spending in order to consolidate the financial recovery. Still, in 2000, four out of ten Russians were reported to live in poverty. The World Bank pointed to growing consumption inequality, which was higher than that of much of the rest of the CIS and Eastern

European transition countries, and stressed the rise of infectious diseases and continued weakness in health services.[38]

Various analysts have described the Russian economy since its recovery from the 1998 Russian financial crisis as still severely challenged—one beset with elements of the oil/natural resource curse, with rampant corruption and minimal industrial diversification, and unreliable legal recourse for enforcing property rights and contract law.[39]

The Russian economy both prospered and sputtered under Putin and Medvedev, largely in response to the rise and fall of global oil and gas prices. Marshall Goldman captured this most effectively in *Petrostate,* and Michael Klare characterized Russia as an "energy juggernaut."[40] The impact of the 2008 global financial crisis curtailed the enthusiasm of most descriptions of Russia as an energy export-fueled global power. Indeed, by April 26, 2016, Senator John McCain, speaking on CNN's State of the Union program could quip: "Russia is a gas station masquerading as a country," as he criticized Putin's Ukraine policy.

There are clearly some good examples of countries using their natural resource riches to strengthen and diversify their economies. Norway is often cited as the best example, where oil income is used to support investment in other parts of the economy and as contributions to a large Government Pension Fund to cover non-oil government deficits through interest earned.

Whether energy income has had a serious impact on the lives of most Russians was still in question as of 2021. Marshall Goldman summarized the key (continuing) problem quite effectively, writing at the height of the boom in Russian oil and gas profits

> ...Russian manufacturers rarely were able to succeed on a purely competitive basis. Russian energy resources were used more as a lifeboat to support a non-market-oriented economy and the country's industrial dinosaurs rather than as a stimulant to growth and the development of a world-class competitive manufacturing complex.[41]

Intent on rebuilding military power and committed to a very expensive construction program for the Winter Olympics in Sochi, the Russian state was not very focused on mass public needs, and there is little reason to expect that a more responsive political system is on the way. The most constructive, forward looking public expenditure one can uncover in recent years was the apparent commitment to consolidating and upgrading the major research universities and the higher educational system more generally. But real expenditures to this end have fluctuated annually along with the variation in oil and gas prices, and there is little evidence in global university rankings to suggest real progress in the creation of world class institutions.

The challenges of simultaneous democratization and economic liberalization are daunting. Aside from arguments that economic reform should be the initial focus, transition in Russia and parts of the FSU was criticized mainly because the social safety net of the state socialist system was dismantled too quickly and completely, without a viable alternative in place. Privatization was pushed for its own sake in some cases, not for reasons of constructive reform. That is, private ownership was seen as the key goal, not competitiveness. There was a one-size-fits-all approach, when in fact diverse models were available with considerable variance in timing and scale. Privatization dominated by insider deals was all too frequent, with disastrous results for average citizens. The rule of law—especially relating to property rights—was absent.

Anders Aslund's case for a strategy of shock therapy or big bang reform as opposed to gradualism was not heeded in Russia. His stress on the importance of minimizing the role of the old state institutions and pushing for speed and comprehensiveness to avoid the "scourge of rent seeking" would prove to be prescient. The rise of the "oligarchs" would demonstrate that those who favored gradual reform most were those who saw ways of making money on the transition to a market economy. Profits in excess of the prevailing level of competition were in effect rents.[42] President Putin regained some control over the economy after 2000 with his campaign to reassert state power, essentially replacing the Yeltsin oligarchs with his own.[43]

Russia, NATO, and the European Union

As noted earlier, the decade after the end of the Cold War radically reshaped the institutional configuration in Europe. Some international institutions disappeared (Warsaw Pact, CMEA/COMECON); others had to adapt themselves to the new reality. The United Nations, OCSE, and the Western European Union have been hampered in their ability to have a serious impact on conflict resolution in Europe. The North Atlantic Treaty Organization (NATO) and the European Union seem to have preserved and partly increased their ability to respond to conflict and promote stability.

A key element of this, of course, is the posture of Russia. Since 2000, Russia has both sought to improve defense cooperation with many of the other successor states of the former Soviet Union, the Commonwealth of Independent States (CIS), establishing a Collective Rapid Reaction Force in May 2001, and to test its ambivalence to a possible role with NATO. NATO enlargement remains a tough issue for many Russians, particularly President Putin in recent years. Some anxiety seemed to be relieved with the establishment of a NATO-Russia Council as a result of the Rome Summit in May of 2002, but the suspicion about Western intentions re-emerged. And Putin found an anti-West (and especially an anti-US) posture to be quite popular with a broad swathe of the Russian population. Playing hard ball with Ukraine on gas pricing and pressuring Ukraine and other successor republics, especially Armenia, not to sign association agreements with the EU became core elements of Russian policy—a policy that may have been counterproductive but is not hard to understand from the Russian traditionalist perspective.

It may well be that the United States and Europe will yet find each other's roles indispensable to global security again, given the Russian invasion of Ukraine in February 2022. The June 2021 G7, EU, and NATO summit meetings orchestrated by the Biden administration did seem to relieve European concerns with some evidence that the United States was back as an important partner on a variety issues, including Climate Change, relations with Russia, and serious cooperation to respond to the emergence of China as a global power. The days of Europeans trying to tie the United States down to a clear commitment to European defense with tripwires and forward basing strategies may still be in question, but continuing security challenges and instability on the periphery of Europe may indicate that European, transatlantic, and Eurasian security require concrete collaboration and diminished rhetoric. The remarkable allied unity over very serious sanctions on Russia, but German reluctance to go further in the near term on energy cuts, reminds us of James J. Sheehan's argument in his important book on the emergence of European civilian states—*Where Have All the Soldiers Gone*. Some of his arguments are still instructive on these prospects:

> Contrary to what some have argued or expected, European states have not disappeared. Indeed, they are more stable and effective than ever before. But they have been transformed: they are now dominated by civilian institutions, focused on civilian goals. European states still have armed forces – just as garrison states had economies – but politically, symbolically, and economically, these military institutions are subordinated to the agencies that do what citizens regard as important: managing the currency, promoting economic growth, providing welfare, and protecting people from life's vicissitudes. The eclipse of the willingness and ability to use violence that was once so central to statehood has created a new kind of European state, firmly rooted in new forms of public and private identity. As a result, the European Union may become a superstate – a super *civilian* state – but not a superpower.[44]

If Sheehan's thesis is true, it seems unlikely that Europe alone will constitute a long-term military counter to Russia's efforts to meddle, pressure, and grab land.

Russia's efforts to use its extensive oil and gas reserves as the principal element of its export profile, as well as to exert political influence, have strengthened European commitment to energy security. This has involved independent national efforts to both expand renewable energy source development and seek alternative suppliers (especially through LNG shipments). European Union programs to reduce carbon emissions add a long-term element to a broader conception of security. Germany's support for the completion of the Nord Stream 2 pipeline to supply natural gas from Vyborg, Russia near St. Petersburg through the Baltic Sea to

Greifswald, Germany (without transit through Ukraine), was hard to understand in this context, and may be indefinitely suspended due to Russia's invasion of Ukraine and the subsequent explosive damage to it.[45]

Nonetheless, European national energy policies were full of contradictions and narrow notions of national versus regional or global interests. Continued reliance on highly polluting domestic coal (especially in Germany), along with limited efforts to reduce dependence on fossil fuels from Russia were unsettling, despite professed security concerns and public commitments to reducing carbon emissions in the context of the EU 2020 Climate and Energy Package, Roadmap 2050, and the 2015 Paris Agreement on Climate Change.[46] European business interests are important, of course, again particularly in Germany where long standing trade of German manufactured products in return for Russian energy and raw material imports have been important since the 1980s.

European Energy Security: Historical Shifts, Current Progress, and Trends

Lester Brown and his colleagues at the Earth Policy Institute argued strongly that *The Great Transition* is clearly underway, suggesting a dramatic shift from fossil fuels to renewable sources of energy, especially solar and wind power. They stress that heavily subsidized fossil fuels are no longer an enviable economic option, particularly when the costs of climate change and health care expenditures (traced to carbon emissions) are taken into account.[47] Similarly, Dieter Helm argued that we are reaching the *Endgame for Fossil Fuels*. There has been clear volatility in the price of oil; it was $147 per barrel in 2008 and then $27 in early 2016. The COVID-19 pandemic reduced demand for fossil fuels substantially for more than a year. The reopening of key industrial and commercial centers in 2022, meant that demand picked up again, and oil prices spiked again due to the Russian invasion of Ukraine. Nonetheless, Helm stressed that the days of sustained high oil prices are over according to his medium to long-term calculations. He criticized the concept of "peak oil," with a complicated argument for: (1) the "end of the commodity super cycle," (2) more effective policies in response to climate change realities and air pollution sensitivity in key fossil fuel consumer countries, and (3) the impact of the growing revolution in technology that will change the nature of the world economy and end the reliance on energy intensive manufacturing.[48] In his view, OPEC countries and Russia face a future of severe budgetary constraint if they continue to fail to diversify their economies. EU member states have made notable efforts to diversify their energy supply sources, but Helm argues that some of the European investment in renewable energy was premature, given the current weakness of energy storage and transmission infrastructure. However, the strength of Europe's educational and technological bases, and the apparent commitment to reducing dependence on fossil fuels, especially that purchased from Russia, given the invasion of Ukraine, will allow it to prosper in the new global energy future that he foresees.[49]

Chapter 3 will look at the emerging European challenges more broadly and comparatively, but a few more words about Germany may be useful now.

Germany is both one of the largest energy consumers in Europe and also one that became one of the most dependent on Russian energy supplies, especially after the end of the Cold War. This chapter concludes with an evaluation of initial EU-level climate change energy policy developments and their impact on German energy policies and outcomes. Detailed comparative assessments of additional European states will push this analysis deeper in Chapter 3.

Table 2.1 depicts the degree of European natural gas dependence on imports from Russia in 2012. The three Baltic countries, Latvia, Lithuania, and Estonia along with Finland, were totally dependent on Russia for their natural gas consumption, while the United Kingdom, Sweden, Spain, Portugal, Ireland, Denmark, and Croatia were at the other end of the spectrum with basically no dependence on Russian natural gas imports. Turkey was 59% dependent; Germany came in at 37%, while France was at 16%. These percentages remained relatively stable during subsequent years until 2022. Those that were at the mercy of Russia did not reduce their dependence, and those that were independent remained so.[50] Germany's dependence continued to increase and was by far the largest total natural gas importer from Russia.

It is clear that the roots established with Ostpolitik to obtain natural gas from Russia took hold more firmly than anyone but the Russians would like. Germany's unification after the fall of the Berlin wall on November 9, 1989, increased that import burden at the same time that it added an enormous public expenditure to try to raise the living standards and circumstances of the "Ossies" to that of their Western compatriots. Germany's public debt burden, which soared with that new and sudden GDR obligation, continued well into the 21st century. But the German economy clearly remains the dominant force within the European Union, and

TABLE 2.1 European Dependence on Russian Natural Gas in 2012 (% of total consumption)

0–10%	*11–30%*	*31–40%*	*50–60%*	*70–90%*	*100%*
Belgium	France	Germany	Austria	Bulgaria Estonia	
Croatia	Italy		Czech. Rep.	Hungary	Finland
Denmark	Luxembourg		Greece	Slovakia	Latvia
Ireland	Romania		Poland		Lithuania
Netherlands	Switzerland		Slovenia		
Portugal	EU-28 average		Türkiye		
Spain					
Sweden					
Spain					
UK					

Source: Eurogas and *The Economist*, April 5, 2014.

Chancellor Merkel played a crucial role in the rescue of the Euro after 2008; this included substantial payments and debt restructuring to help ease the financial crises experienced by several Mediterranean members of the Eurozone, most notably Greece. The importance of Germany's broader economic relationship with Russia after the end of the Cold War is hard to downplay, but this seemed threatened by Russia's actions in Ukraine in 2014, particularly the annexation of Crimea on March 18. But the work on the Nordstream 2 direct pipeline continued largely without interruption until the full scale invasion in 2022.

Natural gas consumption by Germany rose significantly after 1973, more than doubling by 2010. At least 80% of this was imported. Significant amounts were imported from the USSR and subsequently the Russian Federation. Efforts to diversify its sources were somewhat successful, with Russia, Norway, and Denmark contributing 39%, 35%, and 22%, respectively. There was a decline in that supplied by the Netherlands, while imports from Norway and Russia grew as supply sources.[51]

Energy efficiency improvements, such as a district heating strategy, were instituted to reduce total natural gas consumption over the long term, and a renewable energy strategy was stressed in the Government's "Energiewende" (energy transition) legislation in late 2010. The revolution in shale gas fracking technology provided Germany with moderately priced import options; LNG imports from the Middle East, especially Qatar, were also available given declining US imports from the region. The German import options included North American suppliers, as the United States began to convert its LNG port handling facilities from an import to an export orientation. The competitive price pressure on the Russian budget became important; Gazprom had to reduce its prices on gas exports to Germany.[52]

Imports of oil declined moderately in importance relatively for Germany in the Total Primary Energy Supply, from 47% in 1973 to 32% in 2010. Historically, the sources of supply were overwhelmingly external; Germany obtained only 2% of its oil consumption domestically. Russia consistently supplied the largest percentage of Germany's imported oil for several decades. Other suppliers included Saudi Arabia, Kazakhstan, Nigeria, and a range of European countries. Libya's role declined dramatically as a supplier to Germany due to the impact of the devastating civil war in 2011. The mode of transport for oil imports was divided nearly equally between four transnational pipelines and four seaports (Brunsbüttel, Hamburg, Rostock, and Wilhelmshaven). As early as 2012, the German Petroleum Industry Association forecasted that oil consumption would decrease by 14% by the year 2025, due to continued promotion of alternative energy sources—including biofuels, energy taxation levels, and efficiency standards for buildings and cars.[53]

Consistent German consumption of coal, including soft brown lignite—the most polluting form—at first glance is puzzling, given the public declarations of the Government on the need to respond to climate change and thus urgency in moving away from fossil fuels. Some of this relates to the desire to avoid increased dependence on Russian natural gas and oil imports—coal being the only fossil fuel in abundance in Germany. But there was also clearly the impact of domestic political considerations that made it difficult for Chancellor Merkel, burdened with a fragile governing coalition in her final years, to move against the constituencies largely in Eastern Germany, where coal miners made up a large proportion of the labor force. More recently, however, the German coal mining controversy has spread to the West, where, for example, the population of the village of Lűtzerath, near Dűsseldorf, was displaced in late 2022 by a new open pit mine, defended in part by the Government as a response to the impending energy crisis.

What was less clear was why Germany also purchased hard coal from foreign suppliers, including the declining US coal operations in Kentucky and West Virginia. US conversion of coal-fired electricity generation plants to natural gas systems became quite common after 2000, given the abundance of low-cost domestic gas supplies. It is a curious twist that exports to Germany were seen as an important ingredient in former President Trump's promises to rescue the US coal industry.[54]

The role of both nuclear and renewable energy expanded greatly in Germany—each rising from 1% of TPES in 1973 to 11% by 2010. However, nuclear power has been controversial—long the focus of animus by environmental activists. This opposition became most evident when the Green Party rose to participation in the coalition government of Gerhard Schröder in the early 2000s; the Greens pushed through a policy of phasing out Germany's reliance on nuclear energy by 2022. The victory of the center-right CDU in 2005, under Chancellor Angela Merkel, led to an effort to delay the phase-out until 2036, but the nuclear accident in Fukushima, Japan in 2011 brought about an initial decision by Merkel to revert back to the 2022 target. Subsequent discussions and concerns about reliance on Russia for large amounts of oil and gas imports clearly weakened this commitment, particularly after 2014, leading to some continued reliance on nuclear energy in the medium to long term. Disposal of nuclear waste has been an important area of contention. The Gorleben complex has been proposed as a permanent site for storage of spent fuel, but political opposition and some scientific concern about the geology has threatened the plan.

The nuclear accident at Fukushima, Japan in March 2011 seemed to reinforce a dedication to increase Germany's reliance on renewables, especially solar, wind, and biomass options that were already in play. The plan was formalized in the German Energiewende (energy transition) that sought to increase dramatically the percentage of Germany's electricity production from renewables. The target announced in 2011 was a rise to 80% of electricity production, and a goal of 60% of overall energy use by 2050. This would reportedly reduce emissions of greenhouse gases relative to those in 1990, by 70% in 2040 and 80–95% by 2050. Under this plan, Germany would rely on relatively high fixed pricing and tariffs for solar and wind producers for 20 years, as well as preferred access to the electricity grid for them.

One should keep in mind that this strategy was quite expensive, at a time when all of Europe was still emerging from recession following the 2007–2008 global financial crisis; unemployment levels and debt burdens remained high. This of course was accentuated by the more recent challenge of adapting to the costs of the COVID-19 health crisis and the accompanying economic downturn.

Costs to consumers for the conversion project were high, and the new infrastructure required was extensive—some 5,000 miles of new transmission cables had to be laid. The feasibility of this became suspect, given that the EU-wide BANANA [Build Absolutely Nothing Anywhere Near Anyone] syndrome that exerted a strongly negative impact on transmission network expansion. Additional obstacles included the relative power of the Länder in determining infrastructure improvements in the German federal system, and the variance in the suitability of various forms of renewable energy options for specific regions—for example, given the situation of the wind poor Southern regions. Financial incentives led to an increase in (dirty) soft coal consumption. Moreover, the system of subsidizing renewable energy led to a perversion of investment incentives. According to a study by Werner Daldorf, head of the Investment Committee at the German Wind Energy Association, (with 170 commercial wind parks over the course of more than ten years), the return on wind investment over ten years was 25%, compared to expected returns of 60–80%. It can be argued that subsidies have resulted in a misallocation of resources and disappointing investments. Dieter Helm adds to this concern with an argument that much of the German (and EU) investment on renewables was premature, because it was made before the difficult storage and transmission technologies were resolved. The intermittent and low intensity characteristics of solar and wind energy systems in particular limit the efficiency and cost effectiveness of large public expenditures for the immediate future.[55] The distribution of wind energy is indeed a major problem in that most of the production is in the less populated north of Germany, while the population density and industrial concentration is higher in the south, e.g., in Bavaria. Of course, Germany also does not enjoy the consistent sunshine one finds commonly in Southern Europe.

All of this adds to the complexity of the challenge of reducing German dependence on energy imports from Russia, whether this is done to join in on an effort to sanction Russia for its effort to increase influence indeed control over post-Soviet territories like Ukraine, Georgia, Armenia, Moldova, etc., or simply out of prudence to reduce the leverage Russia could theoretically wield over German policy in the future. The financial and political costs of diversification away from fossil fuels are substantial, and while there are demonstrated and emerging alternative sources of fossil fuels, there is economic inertia and potential financial benefit from sticking with Russia as a supplier, given the existing infrastructure. Until recently, Germany had no capacity to receive LNG shipments at any of its ports, though the LNG import capacity at the Dutch "GATE" terminal near Rotterdam was underutilized, operating at less than 20% of capacity.

Key German industrialists beyond the energy sector, e.g., the CEOs of VW, Adidas, ThyssenKrupp, and Siemens, were outspokenly reluctant to risk their lucrative Russian trade and investment operations through European/transatlantic sanctions on Russia after 2014. Siemens CEO Joe Kaeser visited President Putin, after the

Russian annexation of Crimea, apparently to reinforce commitment to the long-term relationship between his firm and Russia. Chancellor Merkel's deputy reportedly chastised Kaeser for his visit and associated statements, but Kaeser was not alone in his opposition to sanctions. The economic dependency between Germany and Russia, begun in the 1970s had grown too strong to endanger lightly. After the 2022 Russian invasion of Ukraine, it appears that German political leaders, especially Chancellor Scholz, were committed to reducing energy dependence on Russia, in cooperation with most other EU members. German industrialists were clearly reluctant to lose Russian markets and investment opportunities, but the sheer scale of the devastation in Ukraine was hard to ignore. We shall explore this prospect more systematically in Chapter 3.

EU Energy and Environment Policy: Development and Dilemmas

The origins of the European Union in some respects can be seen in terms of efforts to address early post-war energy challenges. Two key precursor institutions to the EU were the Economic Coal and Steel Community (ECSC), which helped tie the economic fortunes of France and Germany together in these key sectors as a peace-making and demonstration effort, and the European Atomic Energy Community (EURATOM), which made an initial effort to respond to energy deficits by harnessing nuclear energy for peaceful development and research sharing purposes. Undoubtedly, this interest stemmed from well learned awareness of the importance of energy security that had been so crucial to the outcomes of both World War I and World War II, as well as the need to try to coordinate a European response, given scarce resources and the massive challenges of reconstruction. But the early record also illustrated the dilemmas associated with making serious progress in energy security when the previous experience in intergovernmental collaboration was limited, and the urge toward national autonomy was almost irresistible. Francis McGowan emphasized this in his introduction to an insightful strong group of essays on European national and EU energy policies:

> Most governments have sought to maintain energy policy as a domestic responsibility, intervening either directly or through national firms to maintain some degree of sovereignty. Many regarded energy as being too important to leave to international market forces or to surrender too much to co-ordination within intergovernmental arrangements.[56]

This latter point is understandable given the wide disparity in domestic energy resources and consumption patterns.

Wilfrid L. Kohl provided interesting detail on these challenges in a collection of essays on European national policies by prominent experts, together with his own treatment of EU energy policy in the wake of the so-called Second Oil Crisis. This episode was initiated by the fall of the Shah in Iran in 1978 and was in effect a new oil price crisis, not unlike, at least psychologically, the 1973–1974 rise in prices and embargo by OPEC. European governments responded individually in a crisis atmosphere to attempt to moderate the impact on fragile domestic economies but could not bring themselves to forge a coherent European energy program.[57]

Moreover, it took many years for regional economic integration to develop between governments which initially varied substantially in the emphasis placed on state intervention versus reliance on market forces, as well as levels of industrialization, prosperity, and economic resources available more generally. Nonetheless, some progress was made at the European level, particularly in connection with efforts to reduce carbon emissions and to develop efficient alternative energy systems.

Initial excitement with the signing of the Treaty of Rome in 1957 faded after significant early progress towards a customs union but only modest steps toward a real common market, let alone economic union. Stagnation or "Eurosclerosis" was the common complaint through the 1970s. But the community grew in membership from the original 6 to 9 and then 12 before its explosion in membership after the end of the Cold War to 28 members—with the accession of Croatia in 2013. (Of course, Brexit reduced that back to 27 in January 2020.)

The Single European Act signed in 1986 ushered in a campaign for Europe 1992 to "complete the market." In many respects, this succeeded with the focus on pursuing the four freedoms—free movement of product, capital, people, and services, at least on paper, but the creation of a common energy policy was still not feasible, given national differences (and many more national policies and interests to harmonize). Finon and Surrey, however, described the success subsequent to the implementation of the SEA from EU efforts to liberalize the European energy sector. This was an important advance over the failed attempts by the European Commission to frame a common energy policy in the 1970s and was coupled with inroads to reforming the sector with policies focused primarily on environmental protection.[58]

A series of important steps to make collaborative decision-making more efficient and responsible, at least theoretically, were taken by signing the treaties of Maastricht in 1992, Amsterdam in 1997, Nice in 2001, and Lisbon in 2007. Improved systems of voting, co-decision processes, and increased transparency were important goals, as the EU moved to monetary union and some elements of political union. Nonetheless, angst over the perceived "democratic deficit" and the emergence of a "constitutional crisis" seemed to disrupt progress in the minds of those who sought deeper integration in additional sectors like energy. But this occurred while many others (most notably key British officials) pointed to the value of broader integration geographically as the more important goal.[59]

EU climate change policy in relation to energy security has its roots in the 1991 Kyoto Protocol. This agreement bound the then 15 members of the EU to reduce their collective emissions to 8% below 1990 levels by the years 2008–2012. By 2011, these 15 had lowered their emissions by 14.9%, far over-achieving the target, according to EU data.[60]

The second breakthrough was an EU decision in 2007 to reduce EU emission of greenhouse gases 20% from 1990 levels, raise the share of EU energy consumption from renewables by 20%, and improve EU energy efficiency by 20%, all by 2020. While these targets themselves are not legally binding, the EU passed several binding pieces of legislation to meet the goals, e.g., the creation of a carbon trading system, differentiated emissions reduction and renewable energy targets, and a legal framework for carbon capture and storage.[61] Real progress toward these goals has been mixed. The EU reported reduced emissions between 1990 and 2012 by approximately 18% and increased renewable use up to 13% in 2011, but progress on increasing efficiency remains a great challenge.[62]

It is true that policy deriving from the Euro Crisis and the attendant economic recession cast doubt on the EU's long-term commitment to reducing greenhouse gasses and expanding renewable energy. For example, a 2014 directive from EU Commission President José Manuel Barroso indicated that EU member states were no longer to receive specific guidelines for the development of renewable energy. Of the 63 legally binding targets in EU climate policy covering the period 2010–2050, 62 expired in 2020 or earlier;[63] climate protection in the EU following 2020 seems to have a voluntary character.[64] Further, the EU target for 27% of EU energy from renewables and 40% lower CO_2 emissions by 2030 "won't be split into national targets."[65]

It is clear, however, that the Russian invasion of Ukraine has both strengthened the commitment of many EU members to reduce their dependence on Russian energy and reaffirmed the importance of the EU's Green Agenda for transition to a Net-Zero Carbon future. The related impacts of the Ukraine crisis on global food shortages were stressed most ardently in late May 2022 by Daniel Beasley, Director of the UN's World Food Programme. It was clear that the Russian blockade of grain shipments from Ukrainian ports together with production disruptions in the agricultural sectors of both Russia and Ukraine were catastrophic for developing countries already facing shortages from Covid-19 supply chain constraints and extended drought.[66] These concerns will be explored more thoroughly in subsequent chapters.

Conclusions

So, what does all this mean for energy security and environmental sustainability in Europe and Eurasia? The basic conclusion is that European dependence on natural gas and oil supplies from Russia was long in the making and motivated by a complex mix of domestic economic and political considerations in Europe and in Russia—some still relevant today and not easy to unravel.

There was also a clear historical impact of economic constraints or motives on foreign policy, including the drive for international agreements. We can see this in West Germany's Ostpolitik, and Soviet interest in détente and oil/gas exports in the 1970s and 1980s, but there are also similar connections in the making of the post-cold war world and the dynamics of the late Soviet and post-Soviet transition.

We can see this most notably, perhaps, in Gorbachev's interest in arms reduction given the bankrupt Soviet economy, as well as in European (especially German) trade interests to the East with the collapse of the Soviet Union.

After 2000, Russia intensified its interests in oil and gas exports to Europe and was aggressive in efforts to curtail possible competition. The 2014 Russian pressure on Ukraine (and the initial tepid European response to it) underlined the extent to which this strategy was successful, despite clear statements by the Europeans pledging to begin diversifying away from Russian supplies after the 2006 natural gas price/supply conflict between Russia and Ukraine first began to affect European supply. As we shall see in Chapter 3, the unilateral 2022 Russian invasion of Ukraine added new urgency to European action.

Tekin and Williams have argued that the EU-Turkey connection provided an important challenge to the Russian strategy in relation to the increased oil and gas shipments through Turkey from the Caspian Sea Basin.[67] See Chapter 7 for more discussion of the details of this in connection with Turkey's role in the pipeline diversification strategies of Azerbaijan and Kazakhstan. Russia turned to China in late May 2014 to replace gas exports likely to be lost to Europe in the wake of the Crimea annexation crisis, reportedly at price levels rather lower than those under negotiation for more than ten years. We shall turn to China and its role in the Eurasian Water-Energy-Food nexus in Chapter 6.

The "fracking" revolution for increasing the availability of natural gas supplies to Europe, for example, from the United States, as well as from European domestic shale fields, would seem to loosen dramatically the Russian stranglehold at the same time that European countries are most intent on breaking free from that dependence. As one analyzes the state of the Russian economy in this context, it can be argued that Russia has traded one dysfunctional economic system (Brezhnev's version of central economic planning and burdensome defense spending) for another (Putin's version of an energy cursed system with anemic diversification and stagnant commercial innovation, also with wasteful defense and other unproductive public expenditures).

If we connect this to the multilevel and multisectoral theoretical bargaining framework described in Chapter 1, it is clear that Barry Buzan and his colleagues and their broadened concept of security are on a track that is likely to pay theoretical dividends. There is no question that we need to embrace political, economic, social, and environmental aspects, in addition to the components of "hard" security based on military power. Environmental change and resource scarcity increasingly constitute strong challenge to states and societies concerned by the negative impacts on their ability to achieve broad-based prosperity or even to avoid further economic decline, social turmoil, and forced migration.[68]

Energy security and climate change impacts are crucial to Europe's future immediately and directly as detailed above, but there are long-term, less direct impacts as well. The energy-water nexus and its food security implications, grow in importance to Central Asia, Russia, China, and the Middle East. For example,

Syrian refugees fled a civil war that in important respects was originally induced by water scarcity facing rural agricultural populations. Moreover, the environmental refugees fleeing increasingly arid parts of Africa and Central Eurasia are emerging as political, economic, and security challenges for Europe that seem likely to intensify. The challenges of the water-energy-food nexus in Eurasia touch Europe increasingly. We shall turn to these complicated interconnections in the chapters that follow.

Notes

1 James J. Sheehan, *Where Have All the Soldiers Gone* (New York: Houghton Mifflin, 2008).
2 For a detailed discussion of these negotiations, see Norman A. Graham, "Linkage Politics and Peace Settlements: The Soviet Union and the Germanies in Search of a Peace Settlement, 1969–1974," in *From War to Peace: Essays in Peacemaking and War Termination*, edited by David S. Smith and Robert F. Randle (New York: Columbia University, 1974).
3 See: Wolfgang Wagner, "Towards a New Political Order: German 'Ostpolitik' and the East-West Realignment," *International Journal* XXVII, no. 1 (Winter, 1971–1972): pp. 19–22.
4 *Pravda*, October 25, 1969, p. 1.
5 Graham Hovey, "Why Brandt Looks East," *The New York Times*, October 4, 1971, p. 39.
6 For example, see *The Bulletin*, September 22, 1970.
7 See: Gabriel A. Almond, "The Politics of German Business," in *West German Leadership and Foreign Policy*, ed. Hans Speier and W. Phillips Davison (Evanston, IL: Row, Peterson, and Co., 1957), pp. 235–237; and Pravda, January 30, 1971, p. 2.
8 Pravda, August 25, 1970, p. 5.
9 See: Josef Riedmiller, "Trade with Russia Continues on a Favourable Upturn," *Sueddeutsche Zeitung*, February 16, 1973.
10 *Izvestia*, December 17, 1970, p. 5.
11 Hans Kepper, "Trade Is Key Item in Brandt's Detente Talks with Moscow," Frankfurter Rundschau, March 11, 1974.
12 See: Josef Korbel, *Detente in Europe: Real or Imaginary?* (Princeton: Princeton University Press, 1972), p. 204.
13 Helmut Schmidt, "Why 'Unparalleled' Curbs on Economy Must Be Continued," *The Bulletin* XXI, no. 39 (November 6, 1973): pp. 297–298.
14 C.G. Jacobsen, *Soviet Strategy – Soviet Foreign Policy: Military Considerations Affecting Soviet Policy-Making* (Glasgow: Robert Maclehose and Co., Ltd., The University Press, 1972), p. 208.
15 Wolfgang Leonhard, "The Domestic Politics of Soviet Foreign Policy of the New Soviet Foreign Policy," *Foreign Affairs* 52, no. 1 (October, 1973): pp. 59–74.
16 *Pravda*, October 17, 1971, p. 1.
17 See, for example, Thomas W. Wolfe, *Soviet Interests in SALT: Political, Economic, Bureaucratic and Strategic Contributions and Impediments to Arms Control* (Santa Monica, CA: Rand Corporation, 1971), pp. 5–13.
18 See: Harry Schwartz, "Comecon: The Wide Gap Between Dream and Reality," *The New York Times*, August 8, 1971, Section 4, p. 4; Robert Hunter, *Security in Europe* (Bloomington: Indiana University Press, 1972), pp. 198–201; and Frank D. Holzman, "Soviet Trade and Aid Policies," *Proceedings of the Academy of Political Science* 29, no. 3 (March, 1969): pp. 104–120.
19 *The Bulletin*, December 15, 1970.

20 *Pravda*, February 3, 1970, p. 5; and January 16, 1973, pp. 1.
21 Paul Marer, *Soviet and East European Foreign Trade, 1946–1969* (Bloomington: Indiana University Press, 1972), pp. 44–61.
22 Roger Morgan, *West European Politics Since 1945: The Shaping of the European Community* (New York: Capricorn, 1972), p. 217.
23 *Pravda*, October 11, 1972, p. 4; February 15, 1973, p. 4; and November 4, 1973, p. 3.
24 Jurgen Lorenz, *Luebecker Nachrichten*, October 5, 1973; and Heinz Lathe, *Kieler Nachrichten*, February 27, 1974.
25 Henry Ashby Turner, Jr., *The Two Germanies Since 1945* (New Haven, CT: Yale University Press, 1987), Ch. 4 and 6.
26 Per Högselius, *Red Gas: Russia and the Origins of European Energy Dependence* (London: Palgrave Macmillan, 2013), pp. 218–220.
27 Ibid., pp. 184–186.
28 Ibid., p. 181.
29 Ibid., p. 234.
30 For example, see: Anders Aslund, *Gorbachev's Struggle for Economic Reform, 1985–1988* (New York: Cornell University Press, 1989); Archie Brown, *The Rise and Fall of Communism* (London: Vintage Books, 2009); Zbigniew Brzezinski, *The Grand Failure: The Birth and Death of Communism in the Twentieth Century.* (New York: Collier/Macmillan, 1990); Yegor Gaidar, *Collapse of an Empire: Lessons for Modern Russia* (Washington, DC: The Brookings Institution, 2007); Marshall Goldman, *USSR in Crisis: The Failure of an Economic System* (New York: W.W. Norton & Co., 1983); Norman A. Graham, Folke Lindahl, and Timur Kocaoglu, *Making Russia and Turkey Great Again? Putin and Erdogan in Search of Lost Empires and Autocratic Power* (Lanham, MD: Lexington Books, 2021), Ch. 2; Serhii Plokhy, *The Last Empire: The Final Days of the Soviet Union* (New York: Basic Books, 2014); and Robert Strayer, *Why Did the Soviet Union Collapse: Understanding Historical Change* (Armonk, NY: M.E. Sharpe, 1998).
31 Interview with PBS WGBN Commanding Heights, 10/9/2000. https://www.pbs.org/wgbn/commandingheights/shared/minitext/int-grigoriiyavlinsky.html#2
32 Robert Strayer, *Why Did the Soviet Union Collapse? Understanding Historical Change* (Armonk, NY: M.E. Sharpe, 1998).
33 See: Yegor Gaidar, *Collapse of an Empire: Lessons for Modern Russia* (Washington, DC: The Brookings Institution, 2007), pp. 115–161.
34 See, for example: Marshall I. Goldman, *USSR in Crisis: The Failure of an Economic System* (New York: W.W. Norton, 1983).
35 See: Aslund, *Gorbachev's Struggle for Economic Reform*; and Stephen F. Cohen and Katrina vanden Huevel, *Voices of Glasnost: Interviews with Gorbachev's Reformers* (New York: W.W. Norton & Co., 1989), for detailed descriptions and analysis.
36 For a more detailed discussion of this, see Graham et al., *Making Russia and Turkey Great Again? Putin and Erdogan in Search of Lost Empires and Autocratic Power* (2021), Ch. 4.
37 World Bank, *Poverty in Russia: An Assessment*. Report No. 14110-RU. June 13, 1995, pp. i, ii, 8, 12, 19, 24–30 and 34–36.
38 World Bank, *Russian Federation: Reducing Poverty through Growth and Social Policy Reform.* Report No. 289232-RU. February 24, 2005, pp. i–xiii, 22–25 and Ch. 7 and Ch. 11).
39 See: Marshall I. Goldman, *Petrostate: Putin, Power and the New Russia* (New York: Oxford University Press, 2008), pp. 12–14; Karen Dawisha, *Putin's Kleptocracy: Who Owns Russia?* (New York: Simon & Schuster, 2014); and United Nations Development Programme. *Human Development Report*. New York: 1998, 2005, and 2013.
40 Goldman, 2008, especially ch. 7, and Michael Klare, *Rising Powers, Shrinking Planet: The New Geopolitics of Energy* (New York: Metropolitan Books, 2008).
41 Goldman, 2008, pp. 13–14.

42 Anders Aslund, *How Capitalism Was Built: The Transformation of Central and Eastern Europe, Russia and Central Asia* (Cambridge: Cambridge University Press, 2007), esp. Ch. 2.
43 Graham, Lindahl and Kocaoglu, *Making Russia and Turkey Great Again?*, Ch. 4.
44 Sheehan, *Where Have All the Soldiers Gone*, p. 221.
45 Nord Stream AG.
46 See: Susanne Langsdorf, EU Energy Policy: From the ECSC to the Energy Roadmap 2050.
Berlin: Heinrich Böll Stiftung, December, 2011; and European Union. Commission. *Energy Roadmap 2050*. Brussels, 2013; *2030 Climate and Energy Framework*. Brussels, 2014; and *A Strategy for Competitive, Sustainable, and Secure Energy*. 2015 September 14 https://eur-lex.europa.eu/legal-content/EN/TXT/?uri=legissum:en0024.
47 Brown, et al., 2015, pp. xxii–xiv.
48 Dieter Helm. *Burn Out: The Endgame for Fossil Fuels* (New Haven, CT: Yale University Press, 2017), Ch. 1–3.
49 Ibid., Ch. 8.
50 See, for example, Eurogas, 2015.
51 IEA, Energy Balances of OECD Countries, 2012.
52 IEA, Natural Gas Information, 2012.
53 USEIA, OGEP-Germany, 2012, pp. 6–7; and *Monthly Oil Statistics*, IEA.
54 Bundesverband der Energie- und Wasserwirtschaft.
55 Dieter Helm, *Burn Out*, Ch. 8.
56 Patrick McGowan, *European Energy Policies in a Changing Environment* (New York: Springer, 1996), p. 2.
57 See: Wilfrid L. Kohl, *After the Second Oil Crisis: Energy Policies in Europe, America and Japan* (Lapham, MD: Lexington Books, 1982), especially pp. 186–192.
58 1996, pp. 165–167.
59 See: Norman A. Graham. "A Transatlantic Perspective on the European Union's Constitutional Crisis: Back to the 'Broader before Deeper' Strategy?," in *Europe's Constitutional Crisis: International Perspectives*, eds. Michael O'Neill and Nicolae Paun (Cluj-Napoca, Romania: Editura Fundatiei Pentru Studii Europene, 2007).
60 EU-Climate Action, Europa, January 30, 2014a.
61 EU-Climate Action, Europa, January 30, 2014b.
62 Cerulus, 2013 and EU-Climate Action, Europa, January 30, 2014c.
63 EU European Environment Agency, Europa, January 15, 2014.
64 See, for example: Gregor P. Schmitz, "European Commission Move Away from Climate Protection Goals," *Spiegel Online*. Der Spiegel, January 15, 2014.
65 Sydney Morning Herald, January 24, 2014.
66 CNN, May 19, 2022. See also *The Economist*, May 21, 2022, pp. 11 and 19–22.
67 Ali Tekin and Paul Andrew Williams, *Geo-Politics of the Euro-Asia Energy Nexus: The European Union, Russia and Turkey* (New York: Palgrave Macmillan, 2011).

References

Almond, Gabriel A. "The Politics of German Business," in *West German Leadership and Foreign Policy*, edited by Hans Speier and W. Phillips Davison. Evanston, IL: Row, Peterson, and Co., 1957.

Anders Aslund, *Gorbachev's Struggle for Economic Reform, 1985–1988*. New York: Cornell University Press, 1989.

Aslund, Anders. *How Capitalism Was Built: The Transformation of Central and Eastern Europe, Russia and Central Asia*. Cambridge: Cambridge University Press, 2007.

Brown, Archie. *The Rise and Fall of Communism.* London: Vintage Books, 2009.
Brzezinski, Zbigniew. *The Grand Failure: The Birth and Death of Communism in the Twentieth Century.* New York: Collier/Macmillan, 1990.
Dawisha, Karen. *Putin's Kleptocracy: Who Owns Russia?* New York: Simon & Schuster, 2014.
European Union. Commission. *Energy Roadmap 2050*. Brussels, 2013.
European Union. *2030 Climate and Energy Framework*. Brussels, 2014.
European Union. *A Strategy for Competitive, Sustainable, and Secure Energy*. September 14, 2015. https://eur-lex.europa.eu/legal-content/EN/TXT/?uri=legissum:en0024.
Gaidar, Yegor. *Collapse of an Empire: Lessons for Modern Russia.* Washington, DC: The Brookings Institution, 2007.
Goldman, Marshall I. *USSR in Crisis: The Failure of an Economic System.* New York: W.W. Norton & Co., 1983.
Goldman, Marshall I. *Petrostate: Putin, Power and the New Russia.* New York: Oxford University Press, 2008.
Graham, Norman A. "Linkage Politics and Peace Settlements: The Soviet Union and the Germanies in Search of a Peace Settlement, 1969–1974," in *From War to Peace: Essays in Peacemaking and War Termination*, edited by David S. Smith and Robert F. Randle. New York: Columbia University, 1974.
Graham, Norman A., Folke Lindahl, and Timur Kocaoglu. *Making Russia and Turkey Great Again? Putin and Erdogan in Search of Lost Empires and Autocratic Power.* Lapham, MD: Lexington Books, 2021.
Helm, Dieter. *Burn Out: The Endgame for Fossil Fuels*. New Haven, CT: Yale University Press, 2017.
Högselius, Per. *Red Gas: Russia and the Origins of European Energy Dependence.* London: Palgrave Macmillan, 2013.
Holzman, Frank D. "Soviet Trade and Aid Policies." *Proceedings of the Academy of Political Science* 29, no. 3 (March, 1969): 104–120.
Hunter, Robert. *Security in Europe.* Bloomington: Indiana University Press, 1972.
Jacobsen, C.G., *Soviet Strategy – Soviet Foreign Policy: Military Considerations Affecting Soviet Policy-Making*. Glasgow: Robert Maclehose and Co., Ltd., The University Press, 1972.
Kohl, Wilfrid L. *After the Second Oil Crisis: Energy Policies in Europe, America and Japan.* Lapham, MD: Lexington Books, 1982.
Korbel, Josef. *Detente in Europe: Real or Imaginary?* Princeton: Princeton University Press, 1972.
Langsdorf, Susanne. *EU Energy Policy: From the ECSC to the Energy Roadmap 2050.* Berlin: Heinrich Böll Stiftung, December, 2011.
Leonhard, Wolfgang. "The Domestic Politics of Soviet Foreign Policy of the New Soviet Foreign Policy." *Foreign Affairs* 52, no. 1 (October 1973): 59–74.
McGowan, Patrick. *European Energy Policies in a Changing Environment.* New York: Springer, 1996.
Morgan, Roger. *West European Politics since 1945: The Shaping of the European Community.* New York: Capricorn, 1972.
Plokhy, Serhii. *The Last Empire: The Final Days of the Soviet Union.* New York: Basic Books, 2014.
Sheehan, James J. *Where Have All the Soldiers Gone*. New York: Houghton Mifflin, 2008.
Strayer, Robert. *Why Did the Soviet Union Collapse? Understanding Historical Change*. Armonk, NY: M.E. Sharpe, 1998.

Tekin, Ali, and Paul Andrew Williams. *Geo-Politics of the Euro-Asia Energy Nexus: The European Union, Russia and Turkey*. New York: Palgrave Macmillan, 2011.
Turner, Henry Ashby Jr. *The Two Germanies since 1945*. New Haven, CT: Yale University Press, 1987.
Wagner, Wolfgang. "Towards a New Political Order: German 'Ostpolitik' and the East-West Realignment." *International Journal* XXVII, no. 1 (Winter, 1971–1972), 18–31.
Wolfe, Thomas W. *Soviet Interests in SALT: Political, Economic, Bureaucratic and Strategic Contributions and Impediments to Arms Control.* Santa Monica, CA: Rand Corporation, 1971.
World Bank. *Poverty in Russia: An Assessment.* Report No. 14110-RU. June 13, 1995.
World Bank. *Russian Federation: Reducing Poverty through Growth and Social Policy Reform.* Report No. 289232-RU. February 24, 2005.

3

EUROPE'S ENERGY AND ENVIRONMENTAL CHALLENGES IN THE CONTEXT OF THE GEOPOLITICS OF EURASIA

Is the Russian Invasion of Ukraine a Game Changer?

The simple answer to the above question is a provisional "yes." After years of discussions, proposals, and plans to reduce European reliance on fossil fuels, and indeed efforts focused on reducing imports of oil and natural gas from the Russian Federation, it does appear that the member states of the European Union have new commitment to change their energy systems and dominant fossil fuel supplier. The unambiguous, unprovoked, and inhumane initial invasion, which proved largely unsuccessful in changing the Ukrainian political regime, but caused massive damage and civilian loss of life, led to dramatic European and Atlantic Community unity in response. This included clearcut condemnatory resolutions by the United Nations, the European Union, and many national governments well beyond Europe. Major corporations, including energy firms, acted to withdraw investments and operations in the Russian Federation, and the most comprehensive set of international sanctions targeting Russian assets and leaders was enacted with surprising degrees of severity and substantial global compliance.[1]

Europe and Ukraine 2014

The EU and especially the European members of NATO were serious, at least verbally, in addressing the Russian pressure on Ukraine and the Russian "acquisition" of Crimea in 2014. Selective frozen assets, banking limitations, and travel restrictions were the best that the US and EU members could agree on initially. Serious trade and Foreign Direct Investment (FDI) sanctions were, however, not implemented, despite little constructive response on the part of the Russian authorities to the global opposition. The Obama administration seemed committed to a multilateral response, despite considerable pressure from Republican "Hawks" like John McCain and Lindsey Graham to demonstrate more military strength in response to

DOI: 10.4324/9781003373735-3

the Russian intervention/interference in Ukraine. US troops and equipment were in fact sent to Poland and Romania.

Carl Bildt, former Prime Minister and Foreign Minister of Sweden, argued strongly that the Russian "occupation" of Crimea could not stand, comparing the unmistakable nature of the illegal intervention to Saddam Hussein's invasion of Kuwait in August of 1990. While noting that the Crimea action could not be undone simply or quickly, he argued that Russia had already suffered economically from the initial Western reaction, given the uncertainty investors and traders felt about the viability of the Russian economy and the unpredictability of Putin's behavior. Broad-based support for military action, as in Kuwait, was of course difficult to imagine, especially given that no UN Security Council authorization could be offered without a Russian veto. But the Russian actions in Ukraine did produce serious backlash, even in Scandinavia, where there was for the first-time serious talk in Sweden and Finland about joining NATO.[2]

Background to this conflict includes the 2008 NATO-Russia summit in Bucharest at which it was reported that President Putin suggested that if Ukraine was admitted to NATO, Russia would respond by seeking control over Crimea and parts of Eastern Ukraine. There was no subsequent NATO decision to admit Ukraine, but it seemed clear that Putin had concluded in 2014 that the danger of extended Western influence had become too great in any case, and he in effect delivered on the threat.[3] Some would argue that this was simply a step in Putin's larger plan of "new imperialism," seeking to reconstitute the Former Soviet Union and attain the geopolitical buffer zones that have guided Russian security policy since the emergence of the Russian empire. Marcel Van Herpern, for example, argued that the de-sovereignization of Ukraine would follow from the second Russian war in Chechnya and the effort to exact regime change in Georgia in 2008.[4] A more moderate assessment by Angela Stent stressed four goals in Putin's foreign policy:

1. ensuring that no major international decision can occur without Russian participation;
2. maintain the status quo in the Euro-Atlantic arena;
3. contain and push back Western democracy-promotion efforts in Russia and its neighbors;
4. promote Russia's economic and modernization interests, with access to Western investment and technology.[5]

While perhaps a bit generous in interpreting Putin's moderate intent, Stent's characterization raised troubling questions. For example, if pursued aggressively, do these four goals not cry out for serious unintended consequences and self-defeating impacts? Could Putin have (4) if he pursues the other three without restraint? Does status quo have to mean no new economic relationships for Ukraine and other Eurasian non-members with the EU, as well as no membership of Ukraine in NATO?

The EU established its European Neighborhood Policy in 2004 in an effort to avoid an unfortunate artificial division between the newly admitted members from Eastern Europe and key "neighbors"—countries (especially post-Soviet ones) that remained on the outside. Ukraine became an important target for this bilateral policy and received EU financial support for a series of projects. This was complemented by the EU's multilateral Eastern Partnership.[6] On March 23, 2009, Ukraine and the EU announced a plan for a massive improvement of the natural gas infrastructure in Ukraine. The Russian response was predictably frosty since the announcement was made without consultation with the key natural gas supplier to the country.[7]

The temper changed dramatically with the election of Viktor Yanukovych as President of Ukraine in February of 2010. Clearly pro-Russian in his sentiments, Yanukovych responded to suggestions from Russian President Medvedev quite affirmatively, indicating a willingness to support partial merger of Gazprom and Naftogaz—respectively the Russian and Ukrainian state gas corporations.

The EU drafted an association agreement for the Ukraine in 2012 (along with ones for several other post-Soviet republics). This would require eventual convergence in policies, legislation, and standards to those of the EU, but provided financial support and preferential access to the markets of EU members. Initially, President Yanukovych seemed ready to sign with the EU, but then demurred under pressure from Putin. Russia lobbied for Ukrainian membership in a Eurasian customs union planned with Belarus and Kazakhstan as the other members as of May 29, 2014. (Armenia and Kyrgyzstan subsequently signed in late 2014.) The EU, perhaps ill-advisedly, had indicated that Ukraine must choose between the two potential relationships.

Yanukovych opted for the incipient Eurasian Economic Union (EaEU) over the EU which led to massive protests in Kyiv. This, plus the reaction to the violent government response to the protests, forced Yanukovych to flee to Russia. The association agreement was subsequently signed by his successor, Petro Poroshenko, who took office on June 7, 2014, after a national election. Essentially, the Russian pressure not to sign an association agreement with the EU had the effect of driving much of Ukraine further away from the desired "partnership" with Russia. A pro-Russian Yanukovych government for all of Ukraine was replaced with a pro-Western Poroshenko government, which was then unfortunately hampered by separatist violence in Eastern Ukraine, supported by Russia, and of course the largely diplomatically unrecognized Russian annexation of Crimea in 2014. The region clearly had become more unstable and problematic than at any time since the collapse of the Soviet Union, and Russia suffered dramatically in terms of European, indeed world, opinion and concretely in terms of lost economic advantages and new political and economic burdens.

The EU has long been committed, at least on paper and in leadership speeches, to transforming its energy consumption to renewable, non-fossil fuel sources. The actual performance in this area, however, has been modest in most member states.

Action has not matched the rhetoric. The May 2019 elections to the European Parliament, however, were marked by a substantial shift in the direction of the Green parties in most national delegations. The relative power of the Parliament versus the Commission and Council of Ministers has grown substantially since the 1990s, but it was not clear that this shift in support for environmentally sensitive MPs would translate into more action.

Binding decisions on energy systems remained largely in the purview of national governments. This seemed unlikely to change, particularly in the ongoing climate of discontent with EU overreach, financial weakness, and nationalism.[8] There were substantial discussions about the need to respond to climate change, but little concrete action, prior to Putin's startling invasion of Ukraine. The full-scale invasion was surprising and brutal, involving severe abuse of the civilian populations of cities like Bucha, revealed clearly by war crimes investigators and witnesses after the Russian forces were pushed out.[9] This plus the subsequent almost daily brutal missile and drone strikes on energy, water, and heating infrastructure systems, with apparent indifference to civilian casualties, perhaps together constituted a "game changer" for European action on energy independence.

Europe Responds to February 24, 2022

Incredulous Western policy makers and analysts were treated to aggressive statements by Vladimir Putin in early February 2022. He argued that the United States and NATO were setting a trap for Russia into attacking Ukraine. His reasoning was that the West had not adequately responded to Russian demands to meet its security concerns—essentially to roll back NATO membership, training, and force positioning. Not surprisingly, NATO as a collective defense alliance concerned about the growing threat from Russia was unlikely to respond in a way that would be adequate for Putin.

Russian troops and armored equipment began massing along the Russian border with Eastern Ukraine and in Belarus, north of Kyiv, but many European leaders and analysts still thought a full-scale invasion was unlikely. Even President Zelensky stated publicly that he did not expect an invasion. Others believed that Putin would stop essentially with his new declarations that the separatist dominated provinces of Ukraine—Donetsk and Luhansk—were now independent, indicating that Russian "peacekeeping" forces would move into the territories to secure independence. This would effectively abrogate the Minsk Agreements, but would be both less aggressive and more attainable than a full-scale effort to seize Kyiv and topple the Zelensky regime.[10] After all, Ukraine with its immense territory and a population of about 40 million was an unlikely prospect for successful occupation, especially given the animosity toward Russia that had developed in much of the Ukrainian population since the 2014 (nearly entirely unrecognized) annexation of Crimea. Perhaps Putin really believed that Russian troops would be welcomed as liberators in Kyiv before he apparently surprised even trusted advisors and military commanders with the order to begin a full-scale invasion, but the reality of Ukrainian defense operations and more general resistance was very different. The "march" to Kyiv from Belarus failed miserably, and initial territorial gains in the East and South stagnated or were reversed. Russian military weaknesses in morale, equipment, and disorganization were clearly problematic, and supply chains and tactical operations were not up to the stiff and durable resistance and tactical superiority offered by the Ukrainian forces.[11]

The West labeled this unprovoked and unilateral invasion a "War of Choice" and quickly began to levy the sanctions that President Biden and EU leaders had clearly warned would result from such an action. The stunning severity and broad Western compliance attained was a different story from the rather tepid response to the annexation of Crimea in 2014. On February 25, President Biden announced the Western decision to cut off select Russian banks from the SWIFT system of interbank messaging and account transfers. This was the so-called "nuclear option" never used before as a sanction. This served largely to remove access by Putin and his oligarchs to their massive offshore financial holdings and hampered significantly official financial transactions of the Russian Federation. This initial step was followed by a series of related economic measures that effectively cut Russia off from much of the international economic system and many of Russia's trade and investment partners. Numerous Western corporations ceased operations in Russia, as noted above. Technology transfers and imports were substantially constrained, and key Western energy corporations cut their ties. Indeed, in late February 2022 BP and Shell announced decisions to end participation in Russia with Gazprom and other Russian energy companies and operations. ExxonMobil subsequently announced its pullout from Russia and its Sakhalin-1 operations in the Far East. The Kremlin sought to block this withdrawal with an official decree, but ExxonMobil held fast, challenging the decree with a "notice of difference" which would trigger negotiations. In July, Putin announced the "nationalization" of the Sakhalin-2 project which involved Shell.[12]

Addressing European dependence on Russian energy supplies was more complicated (and more painful), but substantial progress was made in 2022. The EU, UK, and US collaborated quite intensively to eliminate oil and gas imports from Russia, hoping to reduce the income from such sales for Russia, as well as to reduce the ability of Russia to use the European Winter as a "weapon" in late 2022 and early 2023. This of course also supported, at least potentially, the long-standing EU goals to ameliorate contributions to climate change from fossil fuel usage.

After the 2022 Russian invasion of Ukraine, the EU began with new urgency to support programs for increased conversion to renewable energy sources and new sources of energy imports. Some member states suspended efforts to cut back on coal mining and consumption, while others extended nuclear energy generation operations that had been planned for phase out after the Fukushima disaster in Japan. A comprehensive strategy for *EU External Energy Engagement in a Changing*

World was published on May 18, 2022 which discussed these elements in detail and argued for renewed commitment.[13] Importantly, this communication stressed:

> Despite the diversification and internal energy security measures taken since the 2009 gas crisis, Europe is still too dependent on a supplier who is willing to use energy as a weapon. To overcome this vulnerability, the European Commission presented on 8 March 2022 a REPowerEU communication, providing a blueprint to put an end to the imports of fossil energy from Russia before 2030. This requires diversifying the EU energy supply, increasing energy savings and efficiency and accelerating the green energy transition.[14]

The strategy was detailed and ambitious. As suggested in Chapter 2, however, implementation at the national level is essential and variable. Those member states with the greatest dependence on Russia, e.g., Germany, had to make the largest commitments. By the end of 2022, it was still not clear that dramatic adjustments had been made, but there was clear progress.

For example, German Chancellor Scholz seemed largely convinced that it was time to end dependence on Russian energy imports. He signed a deal in late September 2022 for 130,000 cubic meters of LNG to be shipped to Germany by the end of the year from the UAE, following meetings at a "Persian Gulf Summit." He subsequently claimed that Germany was prepared for winter while noting that the German population must prepare for sacrifice.[15]

Nearly a month later, German President Frank-Walter Steinmeier admitted the "bitter failure" of German policy on Russia. He stressed that there must be no peace with Russia as long as Russian troops remain within Ukraine's border. "In the face of evil, goodwill is not enough.... Russia's brutal war of aggression in Ukraine has reduced the European security order to ashes." The speech underlined the shock Vladimir Putin's invasion of Ukraine has inflicted on decades of German foreign policy. Steinmeier was for years one of Europe's most vocal proponents of compromise with Moscow. "A supposed peace that rewards acts like this, a peace that seals Putin's land grabs, is no peace." Settling with Putin "would mean a rule of terror for many people in Ukraine [and] would leave them at the mercy of their Russian occupiers' arbitrary violence", he said.

> Worse yet—a sham peace like this would only increase Putin's hunger. Moldova and Georgia, as well as our NATO partners in the Baltic, live in fear...An unjust peace is no peace—it carries the seed of new wars within it.[16]

Writing in *Foreign Affairs* at the outset of 2023, Chancellor Scholz argued that

> The World is facing a *Zeitenwende:* an epochal tectonic shift, Russia's war of aggression against Ukraine has put an end to an era....Russia's brutal attack

> on Ukraine in February 2022 then ushered in a fundamentally new reality: imperialism had returned to Europe, Russia is using some of the most gruesome military methods of the twentieth century and causing unspeakable suffering in Ukraine.[17]

He went on to sketch out what was necessary to build a stronger Europe while making commitments for Germany to play a more active role and spend approximately $100 billion to better equip the German military. This reflected what he termed a new mindset in German society. Striking a somewhat "glass as half full" posture, he argued that

> Russia's war has not only unified the EU, NATO, and the G-7 in opposition to his aggression; it has also catalyzed changes in economic and energy policy that will hurt Russia in the long run—and give a boost to the vital transition to clean energy that was already underway.[18]

Noting that diversifying Europe's energy supplies was crucial, he admitted that reopening domestic coal mining operations and extending the life of nuclear energy power plants had been necessary, as was suspending the certification of the Nord Stream 2 pipeline with its expected dramatic expansion of gas supplies to Europe. In his words:

> We have learned our lesson: Europe's security relies on diversifying its energy suppliers and routes and on investing in energy independence. In September, the sabotage of the Nord Stream pipelines drove home that message.[19]

So, what have the EU and its members individually done concretely to respond to that lesson learned, and why did it take this startling vicious war of choice to motivate (apparently) serious commitment? Clearly, the vulnerabilities of over-dependence on fossil fuels generally and on excessive imports from Russia had been obvious for many years if not decades, given the overwhelming scientific evidence of the impact of carbon emissions on climate change and the record of Russia previously using energy dependence as a weapon.

Aside from European national decisions to seek alternative sources of energy imports, as illustrated with the Germany-UAE deal noted above, and decisions to suspend efforts to end domestic coal production and reconsider plans to decommission nuclear energy generation plants, there was considerable effort to urge OPEC to ramp up oil production. The UAE reportedly acted to support Western requests in this respect, but Saudi Arabia and other MENA/OPEC producers were not so inclined. This was apparently not so much an effort to support Russian interests as the result of a judgment that oil prices and profits would be affected negatively in the global market. Short-term and economically narrow calculations seemed to win out over the longer-term prospect of weaning Europe off of its dependence on imports from Russia, to the advantage potentially of Middle Eastern producers. Global oil prices were below $50 per barrel for both Europe's Brent crude and the US WTI measures in 2020 and well below $80 per barrel for much of 2021, but they began to rise dramatically after the Russian invasion, reaching a peak of about $125 in March of 2022. The prices remained high through July, but then began to fall. By mid-December, oil prices had declined to below $80 per barrel again.[20]

US and European trade policy toward Russia became more restrictive in March 2022, with President Biden announcing a ban on all imports of oil from Russia to the United States. The United Kingdom and the European Union announced plans to reduce imports of oil and gas by two-thirds by the end of 2022. With bipartisan support from the US Congress, Biden subsequently initiated steps to eliminate Most Favored Nation trading status (Permanent and Normal Trade Relations in the US terminology) for Russia. The move received support from the EU as well, where new tariffs would have a more substantial impact on Russia.

President Biden spoke with Chinese President Xi Jinping by telephone on March 18 to stress that China must not provide new economic support to Russia—in effect to "backfill" the impact of Western economic sanctions. The clear message was that China could then be faced with similar economic sanctions which would likely mean much more to the large export oriented and globalized Chinese economy than was expected to occur with Russia. Xi pushed back, apparently, suggesting that the United States and NATO were responsible for the conditions leading to the Ukraine crisis (adopting the Russian line) and condemned the prospect for widening sanctions.[21] The anti-Chinese government sentiment brewing for some time in the United States and Europe, however, would likely remain strong on "unfair" trade and technology transfer issues. Indeed, intelligence reports in early 2023 that China might be considering supplying Russia with "lethal aid" provoked a new round of warnings to Xi Jinping and his colleagues that the economic consequences of providing weapons or ammunition would be severe.

A key European contribution in response to the Ukraine crisis was the willingness to accept and support millions of Ukrainian refugees, largely women and children, who were fleeing the devastation from Russian bombings and missile attacks. (Men were required to remain in Ukraine, potentially to help resist the invasion, if they were able bodied and of a reasonable age.) On March 24, the Polish Ambassador to the United States reported new legislation passed by the parliament to support better integration of the refugees into Polish society with national IDs and financial support. Polish President Andrzej Duda subsequently urged a fast track to EU membership for Ukraine. Ukrainian refugees were also welcomed throughout much of the EU and in the United States and Canada. There were some complaints as a result, especially from refugee activists, that this posture of "welcoming" was quite distinctive from the postures many European states, most notably Hungary and Poland, took regarding Syrian and North African refugees. One could argue that there were in fact notions of differences in race, ethnicity, and religion at work, but the supportive postures of European governments and peoples could also be attributed to the unambiguous and brutal nature of the Russian invasion. The case for legitimately seeking asylum was hard to deny, and the perceived degree of Russian culpability was high, increasing the sentiment in favor of the Ukrainian women and children. The invasion was indeed a game changer for this and other European policies and actions.

NATO and G-7 heads of state gathered in Brussels on March 24, 2022, to announce an expansion of economic sanctions on Russia and increased aid to Ukraine. This was followed by a US-EU deal to supply LNG to Europe to assist in reducing dependence on Russia for natural gas. The EU President noted that the infrastructure development needed to make this work would take some time but stressed that the construction would be well worth it because it would also be appropriate for transmission of hydrogen when that becomes a viable replacement fuel. Later in the month, President Biden announced that he was ready to tap US strategic petroleum reserves for 1 million barrels per day for six months, to ease the impact of the decline in Russian oil on the world market.[22] On April 3, Lithuania became the first European country to end Russian natural gas imports in response to the invasion.[23]

European, indeed global, animosity toward the Russian invasion of Ukraine was heightened dramatically in early April when atrocities of Russian soldiers were discovered as Ukrainian forces retook the town of Bucha, near Kyiv. Evidence was uncovered of a large number of civilian deaths, many shot in the head with hands bound behind their backs; some of the victims appeared to have been tortured prior to death.[24] The UN General Assembly subsequently voted 93-24 with 58 abstentions to remove Russia from the Human Rights Council.[25] It was not surprising perhaps in this context that public opinion and political pressure rose in Finland and Sweden to move toward NATO membership—a remarkable development, given their long histories of neutrality; but the unambiguous Russian invasion of Ukraine clearly had serious security implications for them and the larger European/Eurasian region. Both Finland and Sweden in fact submitted simultaneous formal applications for membership on May 18.[26] If Putin was really concerned about NATO enlargement since the late 1990s, as he has argued consistently in recent years, his actions in Ukraine must be seen as counterproductive. The dramatic increase in the Russian border territory that now would be contiguous with NATO member territory cannot be an advantage to Russian "security." The invasion would also seem to reaffirm the decisions of former Warsaw Treaty Organization members and Baltic Soviet successor states to campaign vigorously for NATO membership with Article 5 protection, as soon as this was deemed feasible.

A key challenge in the Western response to the Russian invasion was how to reduce the massive income from oil and natural gas exports still available for Putin's military expenditures, beyond simply finding alternative sources and a ramped up commitment to renewables. The EU and United States gradually developed a plan to "cap" Russian oil prices, despite the difficulty of doing so in a global energy market. The G7 formally adopted a price cap of $60 per barrel for Russian oil on

December 3, 2022. EU President Ursula von der Leyen marked the occasion with this statement:

> The G7 and all EU Member States have taken a decision that will hit Russia's revenues even harder and reduce its ability to wage war in Ukraine. It will also help us to stabilize global energy prices benefitting countries across the world who are currently confronted with high oil prices.[27]

This position was strengthened by a new proviso that tankers shipping Russian oil would not be insured or have other maritime services unless the cap was honored even to "third" (non-G7) countries. Western dominance of the world shipping insurance industry put teeth in this requirement. Nonetheless, while innovative and aggressive, the oil price cap was complicated and difficult to enforce, particularly on customers not part of the G7 price cap coalition. A variety of exceptions and contingencies were detailed in the initiative. Whether it would severely reduce Russian oil export income remained to be seen.[28] Putin was clearly upset by the initiative, but it did not seem to dampen his commitment to the invasion. One must remember that economic sanctions in themselves rarely lead to direct change in policy by the target, but simply making an aggressor pay/suffer for the outrageous behavior may nonetheless be worthwhile.

On June 3, the EU formally acted to forbid all Russian oil import shipments by sea while also forbidding insurers from protecting any ships carrying Russian oil. Accordingly, Russia would not easily be able to avoid the European import restrictions by shipping to other markets. All tankers must have insurance coverage for potential damage (especially by Lloyd's of London) and liability (by a European consortium) in order to dock at nearly all ports anywhere. This would greatly hamper Russian export earnings but would also likely increase global oil prices. Later in June, Gazprom cut gas supplies to Europe reputedly in response to the EU's support for Ukraine. Ente Nazionale Idrocarburi (ENI) of Italy reported a 50% cut in its supplies.[29]

The share of gas supply from Russia was still quite high in several countries in 2021—more than 70% in Finland, Bulgaria, Austria, Serbia, and Latvia. Bosnia and Herzegovina, Moldova, and North Macedonia relied on Russia for 100% of their natural gas supply. The dependence on Russia for natural gas by Germany, Poland, Czechia, Slovenia, Hungary, Greece, and Slovakia ranged from 49% to 68%.[30] However, Table 3.1 and Figure 3.1 suggest that EU members, especially Germany, had made significant progress in reducing dependence on Russian energy imports. Indeed, German natural gas imports from Russia dropped from 4.5 billion cubic meters in February 2022 to zero in September 2022, and crude oil imports dropped from 2.4 million tonnes in February 2022 to 3,489 tonnes in January 2023.[31] One could argue that reducing dependence had in fact begun to receive more urgency after Russia annexed Crimea, but this urgency escalated dramatically after February 2022. A rough comparison of the change in dependence on Russian

TABLE 3.1 European Dependence on Russian Energy in 2022 (% of total energy consumption)

0–10%	*11–30%*	*31–40%*	*41–60%*	*61–90%*	*91+%*
Cyprus	Austria	Germany	Finland		Lithuania
France	Belgium	Latvia	Greece		
Ireland	Bulgaria	Poland	Hungary		
Luxembourg	Croatia		Netherlands		
Malta	Czechia		Slovakia		
Portugal	Denmark				
Spain	Estonia				
Sweden	Italy				
Spain	Romania				
	Slovenia				
	EU-28 Average				

Source: Eurostat.

energy sources can be made by contrasting Table 2.1 (2012) in Chapter 2 with Table 3.1 (2022). Clearly the dependence on Russian energy imports was reduced in 2022. No European country remains 100% reliant on Russia for natural gas. Figure 3.1 depicts the success of European energy supply diversification effort.

Progress in reducing European imports of oil and gas from Russia was accompanied fortunately by a somewhat milder than usual winter in 2022–2023. Energy costs were high nonetheless for many Central European households, but the suffering of citizens did not turn out to be as dire as predicted. Several European governments had been remarkably successful in ensuring that their storage capacities for imported fossil fuels were kept replenished.

European governments with port access developed or expanded their capacities for receiving LNG shipments from the United States and the Middle East. Germany opened its first LNG terminal on December 17, 2022, in Wilhelmshaven, reportedly as a direct result of the Russian invasion of Ukraine. The German Government stressed that renewable energy remained a development priority, with four more LNG terminals under construction with completion expected by the end of 2023.[32] LNG shipments from the United States began on January 3, 2023; shipments from the UAE were expected in mid-2023. Deliveries of LNG from Qatar were also expected, but with a somewhat longer lead time.

Beyond Germany, at least 11 additional EU countries have begun to import LNG—Belgium, Croatia, France, Greece, Italy, Lithuania, Malta, the Netherlands, Poland, Portugal, and Spain with a total of 21 operational terminals as of early 2022. This amounted to a total regasification capacity of 160 bcm/year and a total storage capacity of 7.65 million cubic meters.[33]

The Russian invasion of Ukraine also had important impacts on the regional water-energy-food nexus because the Russians blockaded grain and fertilizer shipments

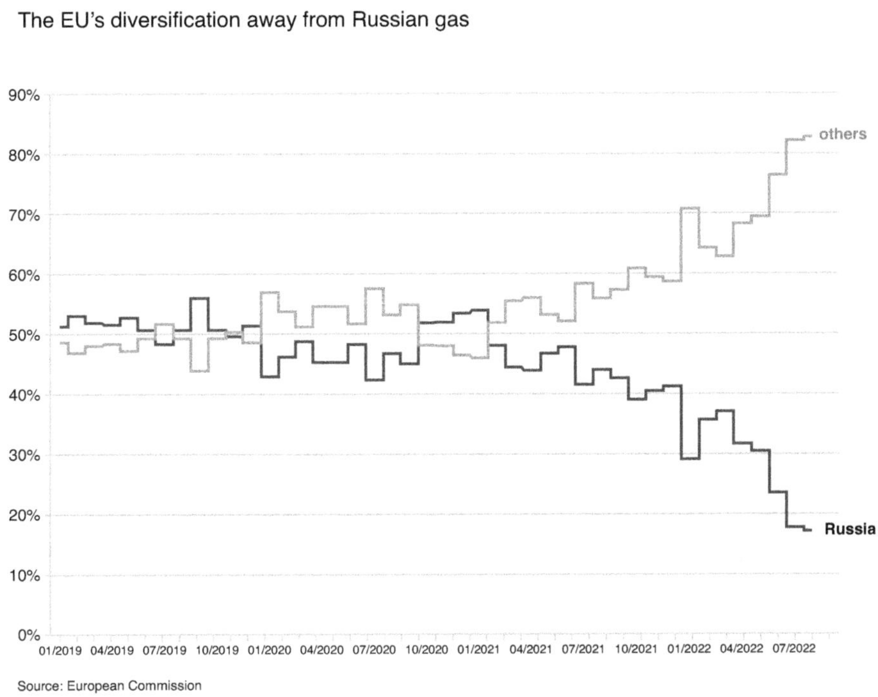

FIGURE 3.1 The EU's Diversification Away from Russian Gas.

Source: European Commission.

from Ukraine's Black Sea ports. Ukraine, often referred to as the "Breadbasket of Europe/Eurasia" is well known for its fertile soil and historically productive agriculture. Daniel Beasley, Director of the UN's World Food Programme, made an urgent plea on May 19 to re-open Ukraine's ports so that grain could be shipped to reduce the impending food shortages plaguing much of the global south—already threatened due to COVID-19 and climate change. Charges and countercharges sought to assess blame for the disruption in grain, sunflower oil, and fertilizer shipments. But there was little doubt that North African and Middle Eastern populations suffered from the decline in imports, including those generally subsidized by the World Food Programme. Ukrainian President Zelensky called the Russian blockade "food terrorism." President Biden condemned Russia for triggering a world food crisis.[34] Russian Foreign Minister Lavrov subsequently promised that the Russian navy would safeguard grain shipments from Odesa harbor if Ukraine would demine it. This of course would negate Ukraine's purpose in mining the region which was to prevent or hinder an amphibious Russian landing at Odesa, in a situation in which Russia largely controlled the Black Sea transport from Ukraine.[35]

UN Secretary General Gutierrez announced that an agreement was finally signed on July 22, 2022 in Istanbul between Russia and Ukraine to permit Ukrainian shipments of grain from Black Sea ports, in an effort to ease global food security challenges. This followed a tour of African capitals by Lavrov to argue that Russia was not in fact provoking food insecurity in Africa. The first Ukrainian ship carrying grain from Odesa was cleared in early August, after inspection in Turkey, to move on to importing countries in Northern Africa. Fits and starts in the grain shipments plagued Ukrainian agribusiness and the importers for much of the rest of 2022.

As the war dragged on into late 2022 and early 2023, European governments sharpened their commitment to assist Ukraine in many ways, despite the Russian efforts to weaponize energy. Germany, not surprisingly given its World War II legacy and its complicated relationship with the USSR and the Russian Federation since, initially showed signs of reluctance to fully commit.[36] Early supplies of equipment were modest and slow in coming, but as noted above, both Chancellor Scholz and President Steinmeier gradually made more and more critical comments on the brutality of the Russian invasion and its unprovoked/illegal nature. The rhetoric began to be matched with shipments and concrete commitments as 2022 ended.

The highlight of this growth in support was the German decision to send 14 Leopard II main battle tanks and permit NATO partners who had purchased their own Leopard II's to ship some to Ukraine in early 2023. The German government was under much pressure from NATO allies and of course President Zelensky to take this step, and a positive commitment was announced only after the United States agreed to transmit M1 Abrams Main Battle Tanks as well. The two systems have similar capabilities and may have an impact on Ukrainian operations to drive out Russian forces from some of their Eastern Ukraine positions. US reluctance to shipping the M1 Abrams tanks was evident reportedly not so much from any serious concern about crossing a red line in escalating NATO support for Ukrainian defense operations, but rather worry about the feasibility of training Ukrainian operational and repair personnel in a timely manner and the suitability of these tanks for the battlefield and transit conditions in Ukraine. The Abrams is much heavier than the Leopard II and requires a specialized fuel mix (not simply diesel fuel). It also may not be able to traverse some Ukrainian roads and bridges as easily as the German tanks and may be harder to keep operational. The show of transatlantic allied unity, however, was important and seemed to reassert the European and North American commitment to support Ukraine despite the costs. US Patriot air defense systems were delivered in 2023 to help protect the capital.

We should stress that the German position on the invasion and on how (much) to support Ukraine is complicated and politically charged. This is not only due to the economic relationship with Russia built up after the Ostpolitik initiatives and the natural gas pipeline diplomacy of the 1970s, as discussed in Chapter 2, but also because of the legacy of broader West German security policies.

One key point to note in this context is the post-War prohibition on permitting West German military forces to operate outside of the borders of the Federal Republic, as well as the limitation on selling or transferring West German-made armaments to "areas of tension." One can see remnants of the initial unwillingness of the German government to grant permission to Poland and other NATO governments to send their own Leopard II tanks to Ukraine in this context. The West German arms transfer agreements frequently had a no-forward-transfer clause so that the original purchases could not in effect violate the "areas of tension" prohibition. Michael Brzoska traced the

> erosion' of this policy despite continuing support for it by the public at large. Brzoska pointed to bureaucratic exigencies and the fact that the German arms industry began to exert a stronger role in shaping the discussions. But the legacy still seems to have an impact on German reticence.[37]

European support for Ukraine was strong throughout the rest of 2022, particularly as the Russian bombings and missile attacks struck city targets with apparent indifference to civilian casualties, and the treatment of civilians in occupied areas appeared criminal to many. Indeed, Poland and Slovakia announced in mid-March of 2023 that they would be sending Russian made fighters to Ukraine—the first NATO members to do so. At that time, there was expectation that the United States would provide replacement fighters, perhaps F-16s to these NATO members in return. In fact, President Biden committed on May 19, 2023 to the Hiroshima G-7 meeting that the United States would begin training Ukrainian fighter pilots on F-16s in Europe. This seems to reflect a growing assessment that the war had progressed to such a stage that enhanced Ukrainian airpower would be important to combat operations. It was also seen as an important step in the support of Ukraine's long-term security needs.

China and India have been key consumers of Russian oil and gas and took a predominantly neutral position, not supporting Western sanctions. India had a long and deep relationship with Russia and enjoyed discounted energy supplies but was not a supporter of the invasion. China also benefitted from reliable oil and gas supplies at good pricing and was not enthusiastic about President Biden's efforts to characterize the conflict as democracy vs. authoritarianism. But Xi Jinping raised questions about the invasion with President Putin and clearly was uncomfortable with this violation of the territorial integrity of a legally recognized independent state.[38]

Xi Jinping visited Moscow late March 2023 for talks on the "no limits" China-Russia strategic relationship. One of the key results, reportedly was an agreement for Russia to supply 236 billion m^3 of natural gas (including LNG) to China by 2030. This would translate into about 30 billion m^3 per year—a sizeable annual export commitment. But if one compares this to the lost exports to the European market, it clearly fails to make up for the lost business. Exports to Europe were about 120 billion m^3 per year. In January 2023, China paid Russia $526 million for 2.7 billion m^3 of natural gas, according to Chinese customs data. This translates into $195 per 1,000 m^3 of gas. In contrast, the natural gas price for the European market was about $700 per 1,000 m^3 of gas in January. Essentially, China received Russian gas with a 72% discount on a substantially smaller total volume.[39]

Prime Minister Modi of India had stressed to Putin that this was not the time for war, given its impact on the world economy. The benefit India enjoyed from lower Russian energy prices was valuable, but the overall crisis was a major distraction—one that could retard India's explosive growth rate and development potential.[40]

Imperialism Versus Territorial Integrity and National Self-Determination

The policies of China, India, and many African states toward the Russian invasion of Ukraine were curious, indeed perhaps startling, in light of their traditional anti-imperialism postures and usual support for territorial integrity. These states mostly did not vote in opposition to the various UN General Assembly resolutions condemning the Russian invasion and calling for the withdrawal of Russian forces from internationally recognized Ukrainian territory, but many did abstain. The Charter of the United Nations clearly recognizes the right of individual or collective self-defense, condemns a state that seeks to change long recognized state borders by force, and recognizes the concept of territorial integrity in no uncertain terms.

For example, Article 51 of the Charter proclaims "Nothing in the present Charter shall impair the inherent right to individual or collective self-defence if an armed attack occurs against a member of the United Nations." There is no question that on February 24, 2022, Russia invaded the territory of Ukraine, an independent member state of the United Nations since at least 1991 and recognized repeatedly in agreements and actions by the Russian Federation. Documentation of this can be seen, for example, in the Budapest Memorandum of December 5, 1994, which (along with similar agreements signed with Belarus and Kazakhstan) provided security assurances from Russia, the United States, and the United Kingdom to these successor states to the USSR if they gave up to Russia the nuclear weapons that had been installed within their borders. The memorandum included accession to the Treaty on the Non-Proliferation of Nuclear Weapons (NPT). It also confirmed Russian, US, and UK agreement to "respect the signatory's independence and sovereignty in the existing borders" and committed them "to refrain from the threat or the use of force against the signatory."[41] Foreign Minister Lavrov's statement in 2016 that Russia never violated the Budapest memorandum because the undertaking was only not to use nuclear weapons in Ukraine is interesting in this regard.[42]

This of course all flies in the face of President Putin's false claims that Ukraine has always been part of Russia and has been captured by an illegal and aggressive neo-Nazi regime with Western support. Putin's article on "The Historical Unity of Russians and Ukrainians" published July 12, 2021, was not convincing, and certainly does not conform well to the insights of highly regarded scholars of Ukraine and its history, like Serhii Plokhy.[43] Putin argued in part:

> I am confident that true sovereignty of Ukraine is possible only in partnership with Russia. Our spiritual, human, and civilizational ties formed for centuries

> and have their origins in the same sources, they have been hardened by common trials, achievements and victories. Our kinship has been transmitted from generation to generation. It is in the hearts and the memory of people living in modern Russia and Ukraine, in the blood ties that unite millions of our families. Together we have always been and will be many times stronger and more successful. For we are one people.[44]

Article 2 of the UN Charter includes the following principles:

> 3 All Members shall settle their international disputes by peaceful means in such a manner that international peace and security, and justice, are not endangered.
> 4 All Members shall refrain in their international relations from the threat or use of force against the territorial integrity or political independence of any state, or in any other manner inconsistent with the Purposes of the United Nations.

The protection of territorial integrity is a key principle of international law and the post-World War II liberal institutional order embodied in the institutions of the United Nations system. In addition to the global consensus enshrined in the UN Charter, a guiding principle of the Organization on Security and Cooperation in Europe (OSCE), established under Article IV of the Helsinki Final Act of 1975, adopted by the USSR and obligated to by the Russian Federation as one of the 57 OSCE members in North America, Europe and Asia, is that "… nation-states should not attempt to promote secessionist movements or to promote border changes in other nation-states, nor impose a border change through the use of force." As Alex Tiersky of the US Helsinki Commission has stressed:

> Russia's illegal occupation of Crimea, initiated in 2014, and its all-out invasion of Ukraine in February 2022, have underlined the extent to which this fundamental principle is under attack in the OSCE region….Members of the Helsinki Commission will continue to actively support the territorial integrity of countries facing Russian aggression and challenge Moscow's claims to illegal changes in internationally-agreed borders.[45]

Does not Ukraine, and indeed all of the other successor states to the USSR and indeed the former members of the Warsaw Treaty Organization in Eastern and Central Europe, have the right to self-determination and independence. As Serhii Plokhy described so well, *The Last Empire* ended in December of 1991. As he stressed:

> …while the lost arms race, economic decline, democratic resurgence, and bankruptcy of communist ideals all contributed to the Soviet implosion, they did not predetermine the disintegration of the Soviet Union. That was caused by the imperial foundations, multiethnic composition, and pseudofederal structure of the Soviet state, features whose importance was fully recognized neither

> by American policy makers in Washington nor by Gorbachev's advisers in Moscow.[46]

In important respects, Putin was clearly intent on recreating that empire, perhaps not fully, but he certainly has no respect for the concept of national self-determination and the territorial integrity of well-established independent states. In subsequent chapters, we shall see the implications and complexities of this apparent return to 19th century style imperialism. Independent states throughout Eurasia potentially face the challenge and must contend with the decline or at least weakness in the defense of these principles in the recent policies of China, India, and many other states, who refuse to condemn Putin's imperialism. This creates worry in the leadership and populations of many post-Soviet and former Warsaw Pact states.

We pose a crucial question now: Will Putin be held accountable for the destruction and murder in Ukraine in violation of international law and the UN Charter? Relatedly, on what basis does he justify this invasion to his general population and his loyal supporters at home and enablers abroad?

Timothy Snyder provided an interesting slant on these questions in *The Road to Unfreedom*. Snyder depicted the Russian leadership throughout the 2014 operations in Ukraine as "…engaged in implausible deniability, telling obvious lies and then daring the Western media to seek the facts."[47] This posture has continued through to the full invasion in 2022 and indeed characterized much of Putin's long speech to the Russian Federal Assembly on February 21, 2023, nearly the one-year anniversary of the initial failed "march" to Kyiv from Belarus.

In Chapter 5, Snyder pointed to the impact of Ivan Ilyin and Vladimir Sorokin, among others, on the information war that played such an effective role in Western inaction in 2014. Sorokin's novel *Almost Zero* published in 2009 revealed his position that "…the only truth was our need for lies, the only freedom our acceptance of this verdict." Snyder framed this in terms of a Russian "politics of eternity" versus the "politics of inevitability" pursued by the West, especially the European Union. Putin's struggle with Western values and models of reform that could not be implemented in his Russia required a strategy of emotional commitment that justified inaction domestically and aggression internationally.[48] Putin's presidential address in February 2023 illustrated this approach:

> One year ago, to protect the people in our historical lands, to ensure the security of our country and to eliminate the threat coming from the neo-Nazi regime that had taken hold in Ukraine after the 2014 coup, it was decided to begin the special military operation. Step by step, carefully and consistently we will deal with the tasks we have at hand.…Let me reiterate that they were the ones who started this war, while we used force and are using it to stop the war.…We are defending human lives and our common home, while the West seeks unlimited power.[49]

The number of lies and mischaracterizations in this speech were numerous and without subtlety or justification. Peter Pomerantsev offered an "insider's view" of the media machine that produces and disseminates these false narratives.[50] M.E. Sarotte, on the other hand, provided a detailed step by step and well-documented recounting of the actual events in post-Cold War Europe and the enlargement of NATO that negates or at least explains the problems with Putin's claims about the legitimacy of the Eastern European drive for NATO and EU membership. The Baltic states, Poland, etc. had good reason in the 1990s to seek collective security sanctioned by the UN Charter. And it was prudent, given subsequent Russia behavior and substantial uncertainty about what Russian instability and ambitions beyond Yeltsin might portend.[51]

On March 3, 2023, US Attorney General, Merrick Garland, visited Kyiv in connection with war crimes and crimes against humanity suggesting that some momentum was building for international legal action. Indeed, on March 17, the International Criminal Court then announced arrest warrants for Putin and Maria Alekseyevna Lvova-Belova, the Russian Commissioner for Children's Rights in the Office of the President of the Russian Federation, because of alleged involvement in the abductions of children from Ukraine. Kremlin spokesman, Dmitry Peskov immediately stated that Russia found the ICC's actions "outrageous and unacceptable." He also stressed that Russia did not recognize the ICC and considered its decision legally void.[52]

Beyond the abduction of children charge, legal scholars and officials have raised the prospect of simply charging Putin formally with the unambiguous crime of "aggression." Russian propaganda, supported by similar narratives offered by China to the contrary, there is little question that Russian actions in Ukraine have constituted unilateral unsanctioned aggression. The appropriate venue for processing such a charge is not obvious, given the existing constraints on existing international legal institutions. There is some argument that a special international tribunal might be convened to hear the case(s) as an alternative. Proponents argue that there is precedent for this, for example, the Nazi war criminals brought before the Nuremburg Trials which were established specifically for the post-World War II context. Could Putin be tried if he was not present in any of these potential venues? Obviously, he is unlikely to participate willingly, and there are then questions about the legality of a trial in absentia. But such a tribunal would sidestep a probable Russian veto in a UN Security Council sanctioned proceeding or a trial before the International Criminal Court to which the Russian Federation is not a signatory. In any case, the ICC arrest warrant at minimum is inconvenient, potentially restricting Putin's international travel. All signatories of the ICC are obligated to enforce an arrest warrant if the "defendant" makes an appearance locally.

In mid-June 2023, leaders of seven African countries traveled to both Kyiv and Moscow in an effort to bring peace to the conflict-torn region. Speaking in Moscow during his meeting with President Putin, President Matamela Cyril Ramaphosa,

President of the Republic of South Africa, was pointed in his view that the invasion must be ended. He argued that the impact of the war on grain shipments and related agricultural trade was devastating to food security in Africa. He insisted that Ukraine's sovereignty must be respected and that the abducted Ukrainian children should be returned to their rightful parents. President Putin interrupted Ramaphosa to state his usual position that the war was legitimate and necessary to remove the neo-Nazi regime installed by a Western-supported "coup," in 2014 and that the children in question were evacuated from Ukraine for their own safety.[53] The African mission was a remarkable departure from the apparent neutrality proffered by much of the global South up to this point, but it was regarded as unlikely to bear fruit, given Putin's "justification" and the Ukrainian position commonly articulated by President Zelensky that an end to the war could only come with a Russian withdrawal of its military forces in Ukrainian territory.

Conclusions

Revisiting the question posed in the title to this chapter, it seems clear that Putin's invasion of Ukraine made an important impact on the energy independence of Europe. This is, however, a complex question with considerable long-term uncertainty. The global oil and gas markets have critical determinants and constraints not so easily manipulated by national policies no matter how coordinated the effort, and the policy environments nationally and regionally are themselves potentially unstable.

One thing seems clear, however, Putin has had to confront a wide range of well-coordinated Western sanctions that leave him with a much more constrained set of resources to fund his ill-conceived war. Questions remain as to how much the Russian economy and industry have been weakened, as well as to what extent the regime has been able to adapt. There are indications that Russia had to dip into its "rainy day" fund of reserves from years of oil and gas profits toward the end of 2022, but some of those reserves were frozen in foreign accounts. Oil income, perhaps at lower prices/profit margins did not disappear, and the government had taken some steps to "sanctions-proof" the economy after the measures that were levied in 2014.[54]

Putin's address to the Federal Assembly in February 2023 was striking both in its lack of detail on the impact of the sanctions or the prospect of specific responses to them. Nonetheless, he warned his oligarchs that it was time to give up on the potential recovery of their frozen assets, foreign real estate, and luxury yachts, and he argued that it was time for Russia to cut off its economic ties with the West and focus on domestic investment and restructuring.

One could argue that on the one hand that this is potentially the much-needed focus on industrial diversification that some Kremlin watchers believed Putin's initial reforms after 2000 might bring, as he clawed back economic power from Yeltsin's oligarchs to fund state programs. He began to speak of the need to reform and strengthen universities and industry beyond oil and gas. On the other hand, much of the tenor of the speech reminded one of the autarkic state-controlled economies of Stalin, Khrushchev, and Brezhnev that were largely isolated from the rest of the world and remarkably backward and inefficient as a result. The evolution of Putin's "kleptocracy" instead, as so well described by Anders Åslund, did not leave Russia in a good position to be prosperous and diversified, if it confronted reduced oil and gas income.[55]

In any case, by design or miscalculation the Russian relationship with Europe has been disrupted, likely beyond repair, without regime change in Moscow. A modest change from within Putin's immediate cronies probably would not be enough to restore the economic partnership. As numerous authors have argued, Putin's advisors and lieutenants seem to have a "Groupthink" mentality, seeing the world in a rather consistent frame of reference, regardless of what contrary information and opinions are available. As Irving Janis might argue, this seemed to entail, at minimum, illusions of invulnerability, inherent morality, and rationalization. Putin, and at least some of his key advisors, seemed to believe that Russia was innocent, victimized, and threatened by the United States and the West at large.[56]

Notes

1 See for example, the detailed narratives and analyses of the first ten months or so of the Russian invasion provided by Luke Harding, *Invasion: The Inside Story of Russia's Bloody War and Ukraine's Fight for Survival* (New York: Vintage Books, 2022); Owen Matthews, *Overreach: The Inside Story of Putin and Russia's War* (New York: Harper Collins, 2022); and Paul D'Anieri, *Ukraine and Russia: From Civilized Divorce to Uncivil War* (Cambridge, UK: Cambridge University Press, second edition, 2023). More than 1,000 companies ceased operations in Russia as of December 2022, according to Jeffrey Sonnenfeld, Chief Executive Leadership Institute, Yale School of Management. https://som.yale.edu/story/2022/over-1000-companies-have-curtailled-operations-russia/

2 Carl Bildt interview with Fareed Zakaria on CNN-GPS, July 13, 2014.

3 For a detailed discussion of the background to Putin's relationship with and ambitions for Ukraine, see D'Anieri, *Ukraine and Russia.*

4 Marcel Van Herpen, *Putin's New Wars: The Rise of Russia's New Imperialism* (Lanham, MD: Rowman and Littlefield, 2014), pp. 1–5 and Ch. 13–16.

5 Angela E. Stent, *The Limits of Partnership: U.S.-Russian Relations in the Twenty-First Century* (Princeton, NJ: Princeton University Pres, 2013), pp. 263–266.

6 See: www.eeas.europa.eu/enp.

7 *Reuters*, March 23, 2009.

8 See: William Drozdiak, *Fractured Continent: Europe's Crisis and the Fate of the West* (New York: W.W. Norton and Co., 2017); and James Kirchick, *The End of Europe: Dictators, Demagogues, and the Coming Dark Age* (New Haven, CT: Yale University Press, 2017).

9 See, for example, the graphic statements made by defecting Russian soldier Nikita Chibrin, as well as the visual evidence of inspectors uncovering mass graves, December 14, 2022. https://www.cnn.com/2022/12/13/europe/russia-defector-war-crimes-with-cmd/index.html.

10 A common view was that Ukraine would be very difficult to occupy in the long term even if an initial invasion was successful. Putin was seen to be clever enough to realize the serious resistance Ukrainians would likely present and the impact that an unambiguous

and unprovoked reaction would not go unnoticed by NATO countries and beyond. See: Graham's interview by Christoph Gunkel in which he admits his surprise at the invasion given these considerations, published in *Der Spiegel*, February 28, 2022.

11 Owen Matthews, *Overreach*; and Michael Schwirtz et al., "How Putin's War in Ukraine Became a Disaster for Russia,' *The New York Times*, December 16, 2022.

12 Justin Jacobs and Max Seddon, *Financial Times*, August 30, 2022.

13 European Union, European Commission. Joint Communication to the European Parliament, The Council, The European Economic and Social Committee and the Committee of the Regions, EU External Energy Engagement in a Changing World. JOIN/2022/23, May 18, 2022.

14 Ibid., pp. 2–3.

15 As reported by the Associated Press, September 26, 2022. https://apnews.com/article/russia-ukraine-germany-prices-cca7e5afd38fbcfc1ad72934e8e59bd9

16 *Financial Times*, October 28, 2022.

17 Olaf Scholz, "The Global *Zeitenwende*: How to Avoid a New Cold War in a Multipolar Era," *Foreign Affairs*, January–February, 2023, pp. 22 and 26.

18 Ibid., p. 30.

19 Ibid., p. 31.

20 https://www.statistica.com/statistics/326017/weekly-crude-oil-prices/

21 *CNN*, March 18, 2022.

22 *Financial Times*, March 31, 2022.

23 *Financial Times*, April 3, 2022.

24 A detailed presentation of the evidence, including satellite images that contradict the Russian claims that the Bucha atrocities somehow were staged by Ukrainian officials is discussed in *BBC News*, April 13, 2022.

25 *UN News*, April 7, 2022.

26 NATO Parliamentary Assembly. https://www.nato-pa.int/content/finland-sweden-accession

27 EU, European Commission, "G7 Agrees Oil Price Cap: Reducing Russia's Revenues, While Keeping Global Energy Markets Stable," *Press Release*, December 3, 2022. https://ec.europa.eu/commission/presscorner/detail/en/IP_22_7468.

28 For details, see: U.S., Department of the Treasury, Limiting Kremlin Revenues and Stabilizing Global Energy Supply with a Price Cap on Russian Oil, December 2, 2022; Oldf Konig, et al. G7 Sets Price Cap for Russian Oil at USD 60 Per Barrel. BakerMcKenzie, December 9, 2022; and Jackie Northam, "What to Know about the Week's New Efforts to Slash Russia's Oil Revenue," *National Public Radio*, December 5, 2022.

29 BBC, June 17, 2022.

30 Statista, 2022.

31 See also: Appune Kerstine, et al., "Germany's Renewables and Efficiency Reforms." *Clean Energy Wire*. Clean Energy Wire, December 7, 2022. https://www.cleanenergywire.org/factsheets/germanys-2002-renewables-and-energy-reforms; "Germany Says It Is No Longer Reliant on Russian Energy." *BBC News*, January 18, 2023. https://www.bbc.com/news/business-64312400; Constanze Steisenmuller, "Putin's War and European Energy Security: A German Perspective on Decoupling from Russian Fossil Fuels." *Brookings*, June 9, 2022. https://www.brookings.edu/testimonies/putins-war-and-european-energy-security-a-german-perspective-on-decoupling-from-russian-fossil-fuels/; and Jake Maly, "The Shifting Dynamics of German Energy Policy," Presentation to UURAF Research Reforum, Michigan State University, April 14, 2023.

32 https://www.dw.com/en/germany-scholz-opens-countrys-first-lng-terminal/a-64134715

33 FTI Consulting, May 10, 2022.

34 CNN, May 19, 2022.

35 See Lavrov's comments at: https://en.topwar.ru/197071-lavrov

36 Foreign Minister Annalena Baerbock referenced this legacy explicitly in February 2022, but she later became a clear advocate for arming Ukraine, pushing the Chancellor beyond his initial comfort level. See: https://www.axios.com/2022/01/21/
37 Michael Brzoska, "The Erosion of Restraint in West German Arms Transfer Policy," *Journal of Peace Research* 26, no. 2 (1989): pp. 165–177. See also Norman A. Graham, "Transatlantic Arms Transfers: The Search for Security, Industrial Cooperation, and Transfer Coordination," in *The Reluctant Supplier: U.S. Decisionmaking for Arms Sales*, eds. Hammond, Louscher, Salomone and Graham (Cambridge, MA: Oelgeschlager, Gunn and Hain, 1983), pp. 203–262.
38 CNN, September 15, 2022.
39 The agreement on future Russian natural gas supplies to China was announced by President Putin and President Xi in joint statements on March 21, 2023. http://en.kremlin.ru/events/president/news/70750.
40 CNN, September 17, 2022.
41 United Nations. Joint Declaration of the Leaders of Ukraine, Russia, the United Kingdom of Great Britain and Northern Ireland, and the United States of America, Budapest. December 5, 1994. https://undocs.org/CD/1285.
42 Russian Embassy in the United Kingdom. "Lavrov: Russia never violated Budapest memorandum." January 27, 2016. See also: Olexander Zadorozhny, "Russian Aggression against Ukraine, the Annexation of the Crimean Peninsula and the 1994 Budapest Memorandum." *European Political and Law Discourse*, 2015.
43 For example, see Serhii Plohky, *The Gates of Europe: A History of Ukraine* (New York: Basic Books, 2015).
44 Vladimir Putin. *On the Historical Unity of Russians and Ukrainians*, July 12, 2021. https://en.kremlin.rus/events/president/news/copy/66181.
45 Alex Tiersky, *U.S. Helsinki Commission, Commission on Security and Cooperation in Europe*, 2016. https://www.csce.gov/issue/territorial-integrity.
46 Serhii Plokhy, *The Last Empire: The Final Days of the Soviet Union* (New York: Basic Books, 2014), p. xvii.
47 Timothy Snyder, *The Road to Unfreedom: Russia, Europe, America* (New York: Tim Duggan Books, 2018), p. 194.
48 Ibid., Ch. 5.
49 Vladimir Putin, *Presidential Address to Federal Assembly*, February 21, 2023, pp. 1–4. http://en.kremlin.ru/events/president/news/70565.
50 Peter Pomerantsev, *Nothing Is True and Everything Is Possible: The Surreal Heart of the New Russia* (New York: Public Affairs, 2014).
51 M. E. Sarotte, *Not One Inch: America, Russia, and the Making of Post-Cold War Stalemate* (New Haven, CT: Yale University Press, 2021).
52 International Criminal Court, *News*, March 17, 2023. https://www.icc-cpi.int/news/statement-prosecutor-karim-kahn-kc-issuance-arrest-warrants-against-president-valadimir-putin. See also: https://www.pbs.org/newshour/world/international-criminal-court-issues-arrest-warrant-for-putin-over-ukraine-war-crimes.
53 BBC World Report, June 17, 2023.
54 David McHugh, *Associated Press*, March 15, 2023.
55 Anders Åslund, *Russia's Crony Capitalism: The Path from Market Economy to Kleptocracy* (New Haven, CT: Yale University Press, 2019).
56 Irving L. Janis, *Groupthink: Psychological Studies of Policy Decisions and Fiascoes*. Second Edition. (Boston, MA: Wadsworth, 1982). For interpretations and illustrations of this at work in the Putin Kremlin, see: Anna Arutunyan, *The Putin Mystique: Inside Russia's Power Cult* (Northampton, MA: Olive Branch Press, 2015); Keir Giles, *Moscow Rules: What Drives Russia to Confront the West* (Lanham, MD: Lexington Books/Rowman & Littlefield, 2023); Norman A. Graham, Folke Lindahl, and Timur Kocaoglu,

Making Russia and Turkey Great Again? Putin and Erdogan in Search of Lost Empires and Autocratic Power (Lanham, MD: Lexington Books/Rowman & Littlefield, 2021), Chapter 4; Henry E. Hale, *Patronal Politics: Eurasian Regime Dynamics in Comparative Perspective* (Cambridge, UK: Cambridge University Press, 2015), Ch. 8; Walter Laqueur, *Putinism: Russia and Its Future with the West* (New York: St. Martin's Press, 2015); Steve Levine, *Putin's Labyrinth: Spies, Murder, and the Dark Heart of the New Russia* (New York: Random House, 2009); Peter Pomerantsev, *Nothing Is True and Everything Is Possible: The Surreal Heart of the New Russia* (New York: Public Affairs, 2014); Brian D. Taylor, *The Code of Putinism* (Oxford University Press, 2018); Mikhail Zygar, *All the Kremlin's Men: Inside the Court of Vladimir Putin* (New York: Public Affairs, 2016).

References

Arutunyan, Anna. *The Putin Mystique: Inside Russia's Power Cult.* Northampton, MA: Olive Branch Press, 2015.

Brzoska, Michael. "The Erosion of Restraint in West German Arms Transfer Policy." *Journal of Peace Research* 26, no. 2 (1989): 165–177.

D'Anieri, Paul. *Ukraine and Russia: From Civilized Divorce to Uncivil War.* Cambridge, UK: Cambridge University Press, second edition, 2023.

Drozdiak, William. *Fractured Continent: Europe's Crisis and the Fate of the West.* New York: W.W. Norton and Co., 2017.

Giles, Keir. *Moscow Rules: What Drives Russia to Confront the West.* Lanham, MD: Lexington Books/Rowman & Littlefield, 2023.

Graham, Norman A. "Transatlantic Arms Transfers: The Search for Security, Industrial Cooperation, and Transfer Coordination," in *The Reluctant Supplier: U.S. Decisionmaking for Arms Sales*, edited by Hammond, Louscher, Salomone and Graham. Cambridge, MA: Oelgeschlager, Gunn and Hain, 1983, pp. 203–262.

Graham, Norman A., Folke Lindahl, and Timur Kocaoglu. *Making Russia and Turkey Great Again? Putin and Erdogan in Search of Lost Empires and Autocratic Power.* Lanham, MD: Lexington Books/Rowman & Littlefield, 2021.

Harding, Luke. *Invasion: The Inside Story of Russia's Bloody War and Ukraine's Fight for Survival.* New York: Vintage Books, 2022.

Kirchick, James. *The End of Europe: Dictators, Demagogues, and the Coming Dark Age.* New Haven, CT: Yale University Press, 2017.

Laqueur, Walter. *Putinism: Russia and Its Future with the West.* New York: St. Martin's Press, 2015.

Levine, Steve. *Putin's Labyrinth: Spies, Murder, and the Dark Heart of the New Russia.* New York: Random House, 2009.

Matthews, Owen. *Overreach: The Inside Story of Putin and Russia's War.* New York: Harper Collins, 2022.

Plohky, Serhii. *The Last Empire: The Final Days of the Soviet Union.* New York: Basic Books, 2014.

Plohky, Serhii. *The Gates of Europe: A History of Ukraine.* New York: Basic Books, 2015.

Pomerantsev, Peter. *Nothing Is True and Everything Is Possible: The Surreal Heart of the New Russia.* New York: Public Affairs, 2014.

Putin, Vladimir. *On the Historical Unity of Russians and Ukrainians.* July 12, 2021. https://en.kremlin.rus/events/president/news/copy/66181.

Sarotte, M.E. *Not One Inch: America, Russia, and the Making of Post-Cold War Stalemate.* New Haven, CT: Yale University Press, 2021.
Scholz, Olaf. "The Global *Zeitenwende*: How to Avoid a New Cold War in a Multipolar Era." *Foreign Affairs*, January–February, 2023.
Stent, Angela E. *The Limits of Partnership: U.S.-Russian Relations in the Twenty-First Century.* Princeton, NJ: Princeton University Pres, 2013.
Taylor, Brian D. *The Code of Putinism.* Oxford University Press, 2018.
Van Herpen, Marcel. *Putin's New Wars: The Rise of Russia's New Imperialism.* Lanham, MD: Rowman and Littlefield, 2014.
Zadorozhny, Olexander. "Russian Aggression against Ukraine, the Annexation of the Crimean Peninsula and the 1994 Budapest Memorandum." *European Political and Law Discourse* 2, no. 6 (2015): 7–13.
Zygar, Mikhail. *All the Kremlin's Men: Inside the Court of Vladimir Putin.* New York: Public Affairs, 2016.

4

LINKING ENERGY, ENVIRONMENT, AND GEOPOLITICS

The Quest for Water, Energy, and Food Security in Central Eurasia

There may be no region in the world where the water-energy-food nexus is more complex and dramatic than Central Asia. The region has been tortured in this respect particularly since the collapse of the Soviet Union. Central Asia moved from centralized control and scarce resource management from Moscow to a situation of independent pursuit of "national" interests that often seemed sharply incompatible (and had potential for serious regional conflict). Countries in Central Asia with abundant water resources are deficient in fossil fuel resources and vice versa, so there is pressure on water-rich countries to increase hydroelectric generation capacity (and some irrigation for marginal agriculture)—moves traditionally opposed by downstream oil- and gas-rich countries which claim substantial water flows for elaborate (and often unsustainable) irrigation efforts of their own.

The South Caucasus successor states (which we add to form Central Eurasia) have a distinctive but no less complex set of challenges for sustainable development and prosperity. The decades long conflict between Armenia and Azerbaijan complicates regional cooperation severely, of course. The Soviet and Russian footprint here, as well as in Georgia, adds to the complexity.

Much of the water and energy sharing imbalance has roots in the legacy of misguided Soviet agricultural and environmental policies that diverted water resources for unsustainable agricultural production goals on the one hand and permitted unfettered industrial pollution of rivers, lakes, and seas on the other. The Soviet water and energy transmission network may have made some sense in Moscow for autarkic economic and heavy industrialization goals under Stalin and his immediate successors, but the damage done to the natural environment and the prospect for sustainable habitation and prosperity in the successor republics to the USSR was appalling. The network of water and energy transmission managed centrally from Moscow in Soviet times disintegrated into decaying infrastructures managed by

DOI: 10.4324/9781003373735-4

national authorities beset with conflicting domestic imperatives and seemingly myopic policy priorities—complicated in most cases by limited economic resources for much of the early post-Soviet period. The ravages of climate change already accentuate the challenges for sustainable development, as the glaciers in China, Kyrgyzstan, and Tajikistan (and China) recede.

There is some promise for sustainable development, however, in the recent urgency placed on economic diversification generally and agricultural development and diversification specifically. This has been encouraged by a sustained period of volatile oil and gas prices globally on the one hand, and new Presidential leadership in Kazakhstan and Uzbekistan on the other.

Central Asia Under the Soviet Union

Central Asia became a key focus for cotton production soon after it was conquered by the Russian Empire; in part, this was encouraged by the constraint on global supply caused by the US Civil War. As Sven Beckert argues in *Empire of Cotton*, the dominant global industry in the late 19th century was textiles, and the growth and processing of cotton was key to the expansion of that industry. The search for land suitable for cotton cultivation was intense, and the fact that there had been modest cotton production in Central Asia for some time by indigenous peoples made it seem a likely candidate for Russian expansion.[1]

The region was not ideal, however, due to its arid nature. Substantial irrigation would be required, given the limited annual rainfall in the region. The seeds largely came from the United States, and the expansion of production through the construction of an elaborate irrigation system began. Replacing imports of "white gold" from the United States was important to Russia, given its own burgeoning textile industry. Cotton produced in Central Asia became an important export crop as well. This approach worked well until the Russian revolution. Anti-Bolshevik rebels tried to use the chaos of the revolution to promote independence movements in the Central Asian regions of the Russian Empire. Some irrigation networks were destroyed, and agricultural production diminished generally, leading to serious famine.[2]

After the "Reds" prevailed in the Russian civil war, the irrigation system was rebuilt and expanded. Migration to and colonization of Central Asia and Siberia were important features, in an effort to control the local populations and tie them and the migrants to the agricultural development effort and increase in cultivated land; this continued what McCauley called the "Trek to the East."[3]

Stalin's efforts to collectivize agriculture were felt severely in Central Asia. The massive famine that accompanied it in Ukraine, afflicted Kazakhstan harshly as well. Part of the challenge was to "persuade" the nomadic populations of Central Asia, particularly Kazakhstan, to settle and convert their lives to collectivized production.[4] Stalin's priority to focus on rapid heavy industrialization to catch up with Western industrial countries, "or they will destroy us" over-ruled the more modest and diversified approach advocated by senior Politburo member N.I. Bukharin, for example, to avoid a confrontation with the rural majority.[5]

Agricultural reforms were necessary, as Zhores Medvedev stressed, but a less collectivized and "disciplinarian" approach probably would have led to more production:

> The growth of productivity of arable land—slow but visible in 1906–1916 and in 1923–1928—stagnated, and the yields of the main cereal crops did not grow for the next twenty-five years. The total grain harvest, which was 86 million tons in 1913 remained at the same level until 1952–1954...despite a significant population growth.... The political price of Stalin's collectivization was even higher. The methods of terror used against millions of kulaks and other peasants and against agricultural experts and the tragedy of the famine of 1932–33 created a serious rift in the relationship between the government and the masses. The Party was dis-oriented. An official purge of the whole Party began in December 1932, by which time the most able and moderate leaders had been expelled. This purge created the foundation for the cult of Stalin and his ultimate dictatorship. The terror of 1936–1938 has many links with the terror of 1929–1933.[6]

The Soviets turned their attention to the "transformation of nature" for expanded agricultural production in Central Asia. The Great Fergana Canal was constructed in 1939 in just 45 days! Remarkably, it was 270 kilometers long, diverting water from the Syr Darya River to irrigate the fertile Fergana Valley primarily for cotton production; but it was also justified as a means to improve rural access to reliable water supply and to increase production of vegetables and wheat. It reportedly was achieved by employing more than 160,000 Uzbek and Tajik collective farm workers. As prominent Russian historian, Lewis Siegelbaum, noted: "...the canal also set in motion forces that would result in the desiccation of the Aral Sea, one of the great ecological disasters of the second half of the twentieth century."[7]

The pace of constructing the irrigation network picked up after World War II. This included an effort to tame the dry winds coming off the desert areas by an extensive program of afforestation. The trees tended to wilt in the hot, dry areas, however, so four large irrigation projects were announced in 1950. The fourth of these was based in Central Asia and focused on diverting water from the Amu-Darya River to irrigate desert areas along the lower river and beyond as far as the south-eastern coast of the Caspian Sea through the long Turkmen and Karakum canals. Six million hectares were converted to cropping in Turkmenistan, and another 22 million for expected additional grazing land created from semi-desert areas.[8] Irrigation was not new to the region, but the scale was larger and perhaps the target less rational than the irrigation used in Central Asia before it was conquered by the Russians (given the clear stronger focus on cotton production). Cotton production expanded substantially as a consequence of the irrigation and the ready access to cheap (forced) labor. The Aral Sea gradually diminished from the reduced water flow. Central management from Moscow made the production system somewhat "viable" for the upstream and downstream regions, assuming the disappearance of the Aral Sea was acceptable.

The Irtysh-Karaganda canal in Northern Kazakhstan (451 kilometers long) was started in 1962 and completed in 1974. It included 22 pumping stations to raise the water more than 450 meters during the course of the transit. This meant substantial energy consumption, but electricity was cheap in the USSR, given its vast oil, gas and coal reserves. The Irtysh River originates in Russia but also passes through China, which indicated plans to divert some of the water for its own irrigation and power generation purposes.[9]

The Soviet leadership had ambitions to transform nature with even more dramatic river diversion plans. In October of 1984, the Central Committee of the CPSU approved a long-term program of irrigation to overcome the continued stagnation in Soviet agricultural production. This included a dubious plan to divert Siberian rivers to Central Asia through new canal construction by 2000. The Siberian river diversion program was canceled in 1986, however. Questions were raised about the ecological impacts of such a massive project, but it may simply have been a question of the massive costs involved—at a time when the Soviet economy was seriously weakening.[10]

Expansion of Soviet grain production in Central Asia was an important long-term goal, particularly after the death of Stalin. Nikita Khrushchev in fact used the post-World War II agricultural production crisis to support his struggle against other Politburo contenders, especially G.M. Malenkov, to succeed Stalin. Most notable in this was his launch of a "Virgin Lands Campaign," first outlined in a memo on January 22, 1954. This was to involve "... the cultivation of 13 million hectares of virgin land in Kazakhstan, East Siberia, the Volga basin and the North Caucasus."[11] As Peter R. Craumer, argued, there were serious problems with the campaign:

> During the early years of the program, the Soviets used methods inappropriate for such dry conditions, partly because the large numbers of agricultural workers who went there from other parts of the country were inexperienced in dryland farming and partly because of the lack of appropriate machinery. In addition, there were strict planting quotas, which effectively prevented the use of summer fallow, and whole-sale meddling in agronomic decisions by party leaders—especially Khrushchev—who had no knowledge of proper dryland farming methods.[12]

William Taubman documented Khrushchev's agricultural reform challenges in exquisite detail, stressing his unwillingness to address the obvious faults in the larger ideological system. Khrushchev's mercurial ranting at the continual production shortfalls did little to improve the dynamics of the problems.[13] Martha Olcott, however, has described the details of the Kazakh organization of the effort and its positive impact on the agricultural and economic development in the region.[14]

As has been commonly argued, the characteristics of the Soviet centrally planned, command economy, together with the excessive defense spending under Brezhnev, left the economy in desperate shape by the time Gorbachev came to power. His policies of *glasnost* and *perestroika* were largely too little, too late to save the sinking ship. An enormous and growing technological gap had opened up with the West, given the lack of scope and incentives for innovation. Lacking competitive commercial drive, the system could not respond to demands for consumer products, including quality food. Stefan Hedlund discussed this in detail, noting the dramatic impact on peasants under the general system, describing agriculture as "A Black Hole in the Soviet Economy," and stressed his pessimism about Gorbachev's modest privatization reforms as a "Pyrrhic victory."[15] D. Gale Johnson and Karen McConnell Brooks also assessed the prospects for agricultural reform in the 1980s, stressing that climatic factors were largely not determinative in explaining the poor performance of Soviet agriculture. They placed considerable emphasis instead on the ineffectiveness of the Soviet decision-making system and the missteps of key policy makers.[16]

The continuing focus on raw material exports (especially oil and gas) to earn hard currency for necessary imports was debilitating, particularly when oil and gas prices were low. This, of course, made it hard to finance the needed grain imports, given persistent poor harvests. The virgin land program of "extensive" agriculture was largely unsuccessful. But it was not clear that a focus on intensification would have worked better without appropriate technology and better trained personnel. This, in turn, was not feasible without basic reforms to the political and economic system. Gorbachev spent some time in charge of agricultural reform under Brezhnev, experimenting with some options on the pricing system. But the ideological opposition and the failure to provide real incentives to encourage more productivity left the Soviet agricultural sector performing rather less successfully than some of the Eastern European state socialist systems.[17]

Excessive centralization, an inappropriate price system, and a defective set of incentives for farm managers and workers plagued Soviet agriculture well into the 1980s.[18] Although Gorbachev did visit China after assuming leadership of the Soviet state, he seemed not to grasp the importance of the relatively simple feature of Deng Xiao Ping's agricultural reforms in the late 1970s (e.g., allowing farmers to keep production over the state quotas and use the sales for productive investments). Deng's reforms along these lines had begun the remarkable success story of the rise in China's agricultural productivity and subsequent miraculous economic growth.

There was some hope that the collapse of the Soviet Union might enable agricultural reform—and it did in some of the successor states. Central Asia, however, did not take agricultural restructuring seriously in the early post-independence years. The region was beset with several key areas of environmental distress, and the transition reform process was far from optimal.

Dislocation from the Collapse of the Soviet Union

It was hardly surprising that the dissolution of the Soviet Union in December of 1991 led to a period of dislocation and chaos in Central Eurasia. The search for stable political regimes and economies in the wake of sudden independence was unchartered territory for the successor states, especially those where much of the leadership initially seemed unsure of the desirability of independence (from the well understood processes and expectations under the Soviet umbrella). Several crucial challenges arose immediately.

First, the collapse of the Soviet economy was dramatic and far reaching. With the subsequent formal dissolution of the Soviet Union as a political entity, the tasks for each successor state in the dual challenges of building both new political institutions and a productive economic system were immense. The common Western advice of democratization and economic liberalization was largely not adopted to any significant degree among the successor states, save in the Baltic republics and perhaps in Georgia beginning with Mikheil Saakashvili's "Rose Revolution." For the rest, particularly in Central Asia, when some elements of the suggested reforms were adopted, they were only modestly effective. The literature explaining this story (these national stories) is detailed and interesting, but too large to address here systematically.[19]

Second, the long-term negative environmental impacts of policies begun under the Soviet regime affected all Central Asian countries to some extent—unchecked pollution, water wastage, destroying habitat and reducing, even eliminating formerly productive fisheries. These actions were long in the making and not easily remediated. The most striking fact, however, is that until recently, none of the successor regimes in Central Asia sought to address, on the one hand, the harmful effects of wasteful irrigation to grow inappropriate crops like cotton and rice in excessively arid regions, and on the other to pursue less expansive hydroelectric generation strategies and polluting practices that had disrupted human and animal/fish habitat so dramatically. The desertification of the Aral Sea in Kazakhstan and Uzbekistan is a well-publicized, well-studied case, and the most striking example of this tragic set of policies. The elimination of this formerly productive fishery had obvious direct impacts on the livelihoods of the fishers, but the nearly complete destruction of the Aralsk/Kzyl-Orda and Muynak/Karakalpakstan regions in Kazakhstan and Uzbekistan respectively, and their broader populations, through the secondary impact of soil encrusted with salt and poisoned by pesticide runoff and then spread through the air in the common regional dust storms, was nothing short of devastating. Breckle et al. provided a detailed analysis and assessment of the Aral Sea disaster. Micklin, Aladin, and Plotnikov offered a similar assessment, but included a variety of sectoral analyses, offering some hope for recent efforts at rehabilitation, at least on the Kazakh side of the region.[20] Uzbek President Islam Karimov reflected on the disaster of the Aral Sea and its causes in some detail in 1998. By then, it was likely too late to stop much of the deterioration, but his analysis did suggest some real awareness of the situation, if not an actionable program of near-term remedy, given the regional dynamics.[21]

It remains clear, however, that newly independent governments mostly chose not to turn their backs on the Soviet policies and system that had wreaked havoc on fish habitat and the larger environment.[22] The Soviet imperial policies of autarky and self-sufficiency (especially behind the cotton monoculture focus in Central Asia) were no longer in play as a political justification for bad economics and agriculture, especially in the increasingly globalized economy of the post-Cold War world. The extent of environmental degradation was dramatic.

One can argue that the political leadership in each country was distracted by more pressing governance and economic development challenges in the early years of independence—indeed for Tajikistan, civil war raged on during the early years. More pedestrian, but nonetheless evident, was the fact that most of the regimes were more focused on other elements of economic and development policies. Environmental policy was well down the list of reform priorities, as was agricultural reform and rural development more generally. The political change from life as part of the Soviet Union was massive, but many assessments suggested that the centralized administrative system and structures were remarkably similar to the Soviet system, for many years after independence. This was often cited as an obstacle to real economic progress.[23]

The third set of factors that explain the environmental crisis in Central Eurasia has to do with the crucial water-energy-food nexus operative in much of the region and the inability to find a strategy to share water and energy resources equitably. Countries with abundant water resources (Kyrgyzstan and Tajikistan) were deficient in fossil fuel resources, while oil- and gas-rich countries (Kazakhstan, Turkmenistan, and Uzbekistan) were downstream from and "dependent" on the water-rich countries for agricultural irrigation (and fisheries). The pressure on water-rich countries to increase hydroelectric generation capacity is clear. But such action was often opposed by downstream oil- and gas-rich countries, especially in Central Asia, which claim a right to substantial water flows for elaborate irrigation efforts, especially in support of the cotton monoculture. Indeed, the International Crisis Group (ICG) argued as early as 2002 that

> Competition for water is increasing in Central Asia at an alarming rate, adding tension to what is already an uneasy region….An annual cycle of disputes has developed between the three downstream countries—Kazakhstan, Turkmenistan and Uzbekistan—that are all heavy consumers of water for growing cotton, and the upstream nations –Kyrgyzstan and Tajikistan. The downstream countries require more water for their growing agricultural sectors and rising populations, while the economically weaker upstream countries are trying to win more control over their resources and want to use more water for electricity generation and farming….Control and enforcement mechanisms no longer function, and the various countries now often accuse each other of exceeding quotas.

> Turkmenistan is using too much water to the detriment of Uzbekistan, which in turn has been accused by Kazakhstan of taking more than its share. Kyrgyzstan and Tajikistan say that the three downstream countries are all exceeding quotas. Even within Uzbekistan, provinces have accused one another of using too much water....Some of the most serious tensions have centered around barter agreements and payments. The upstream countries trade water to Uzbekistan and Kazakhstan for energy in the form of gas, coal or power. Since energy deliveries have been unreliable, Kyrgyzstan has responded by releasing more water through its hydropower dam in winter, which results in downstream flooding and less water for summer irrigation. Attempts by Kyrgyzstan to demand payment for water have been resisted by the downstream countries.[24]

It is clear that countries with abundant water resources—Kyrgyzstan and Tajikistan—have not been able generally to monetize this resource, at least not in comparison with what the oil- and gas-rich countries have been able to do. They generally needed to switch to electricity for winter heating as the market prices for fossil fuels rose dramatically from what they were under the Soviet system. Indeed, schemes to compensate water draw down with energy resource transfers, while the subject of serious negotiation and some agreements, most notably between Kazakhstan and Kyrgyzstan on the one hand and Uzbekistan and Tajikistan on the other, have not worked well.[25]

Indeed, former Uzbek President Islam Karimov suggested possible recourse to military action in response to potential cuts in water flows, for example, if the Tajiks completed construction of the Rogun dam.[26] The "energy-water" nexus was clearly a challenge, as Tajikistan and Kyrgyzstan on the one hand, and Uzbekistan and Kazakhstan on the other, have seemed to court serious conflict over the hydroelectric generation plans of the former, in competition with the downstream irrigation demands of the latter. Irrigation needs are problematic due to aging and poorly maintained structures, harsh climate, and soil quality deficiencies, but until recently they have been driven by the surprising survivability of the legacy of Soviet cotton monoculture in the region.[27]

The International Crisis Group and the World Bank both suggested systems for sharing and domestic reform in an effort to resolve the regional conflicts of interest and ambitions, in 2002 and 2004, respectively. As each country sought advantage and independent solutions to their energy-water nexus constraints, a regional solution seemed out of reach.

Tajikistan sought to control the water and hydropower resources by constructing the high Rogun Dam, over the objections of Uzbekistan. Turkmenistan planned creation of the Golden Century Lake in the Karakum Desert, ostensibly from agricultural irrigation runoff, but with the widespread expectation that the Amu Darya would be drawn upon to replace the shortfall. Uzbekistan revived interest in the diversion of Siberian rivers (Irtysh and Ob) to meet its water needs (potentially capitalizing on Russian interest in more influence in the region). None of these options were supported by the foreign assessments. Indeed, emphasis was placed on revising the existing sharing and barter agreements based on realistic and fair economic and environmental impact assessments of the five participants.[28]

More recently, Andrea Zinzani examined two key water basins in Kazakhstan and Uzbekistan in terms of the implementation of Integrated Water Resources Management (IWRM) measures supported by international donors, including water user associations and market pricing mechanisms. His fieldwork enabled him to provide a detailed exposition of how the integrated effort ran into severe implementation problems, as it confronted serious social, political, economic, and bureaucratic constraints, especially at the local level. His concluding assessment, argued in part:

> ...the IWRM has a really strong, though somewhat hidden, ideological and political aim to change the existing political structures throughout the world, as evinced from their support of rolling back the state, to the privatization processes and the bank investments actions, and the introduction of monetary mechanisms. The evidence that emerges from the Uzbek and Kazakh case studies shows how the political nature and related implications of the IWRM implementation are strong, and also demonstrates how they aim to shape the governmental and social structures. However, it seems quite clear that in both countries this rationale, so far, has only been able to scratch the surface of a long-established cultural and political system which is quite the opposite of the one supported by the IWRM and the implementing agencies.[29]

Oil and Gas Development as a Lifeline for Kazakhstan and Turkmenistan

The feasibility of a fair water and energy resource sharing and barter arrangement had to be connected with the realistic prospects for oil and gas development and export performance in the three downstream, energy-rich countries: Kazakhstan, Turkmenistan, and Uzbekistan. Fortunately, the circumstances of the global energy market, the support of Western governments and China, the domestic preoccupation of Russia under Yeltsin, and the clever opportunism of Central Eurasian governmental elites came together to facilitate energy independence and strong export relationships, at least until the global price of oil and gas fell dramatically after 2008. The volatility of global oil prices remains a challenge for long-term development planning, of course, and the long-term commitments globally to reduce reliance on fossil fuels adds complexity to the decision process.

Kazakhstan's initially successful energy strategy was captured well in Ariel Cohen's 2008 book: *Kazakhstan: The Road to Independence –Energy Policy and the Birth of a Nation,* which placed stress on the delicate balancing act that President Nazarbayev needed to play in pursuing oil and gas field development and independent marketing and transportation options. Nazarbayev took steps to ensure that Kazakhstan would be viewed as an energy partner to Russia, not a competitor, as it divided its exports through the Atyrau-Samara pipeline to the Russian distribution network, the operations of the Caspian Pipeline Consortium (also through Russia to Europe via the marine terminal at Novorossiysk on Russia's Black Sea coast) and via tankers across the Caspian to Baku, Azerbaijan and Neka, Iran, and eventually via the Atasu-Alashankou pipeline to China.[30]

Steve Levine's book, *The Oil and the Glory: The Pursuit of Oil and Glory on the Caspian Sea,* described the intricate negotiations that Presidents Nazarbayev and Heydar Aliyev (of Azerbaijan) undertook to finance development of oil and gas fields with investment from Western energy companies and to develop the option to ship directly to Europe, through the planned Baku-Tbilisi-Ceyhan pipeline. President Bill Clinton played an important role in supporting this effort, and President Yeltsin perhaps made it possible, by his disinterest, given his preoccupation with domestic political and economic problems in Moscow. The BTC pipeline provided its first load of petroleum to an Italian tanker in May 2006, but it was only Azerbaijani oil, as the Kazakh President elected to keep the initial Kazakh production and pipelines focused on the Russian network at that point.[31]

Levine pointed out that while Nazarbayev thus "failed to build geopolitical leverage" initially, he did construct a magnificent new capital, Astana, and noted:

> Grandiosity aside, by 2007 he had led Kazakhstan to the brink of vast wealth. Everywhere was a building boom. A fairly large and affluent middle class was expanding. The economy had grown by nearly 10 percent a year since 2000, per capita income had more than tripled to about $5,000 a year, and the percentage of the population in extreme poverty had fallen to 10 percent. Some $17.5 billion of the country's revenue was deposited in a national rainy day fund, and any future acceleration in crude oil production would build it further.[32]

A parallel pipeline network for natural gas shipments from Kazakhstan near the Uzbek border to Khorgos, Xinjiang, China was over 2,000 kilometers in response to a 2007 agreement signed by President Nazarbayev and Chinese President Hu Jintao. The first substantial shipment began in October 2017, adding to the annual natural gas exports to other customers of 13.7 billion cu m in an effort to strengthen the Kazakh economy. CNPC took delivery.[33]

Tables 4.1 and 4.2 trace Central Asian oil and natural gas production and exports. Kazakhstan clearly leads the way with substantial oil and gas production and exports throughout its post-independence history. Even in 1991, Kazakhstan produced 26.6 million tons of oil and exported 20.9 million tons of it. By 2015, production had expanded to 80.2 million tons, and exports grew to 63.6 million tons. Production increased to 86 million tons in 2021. Turkmenistan and Uzbekistan produced 5.4 million tons and 2.8 million tons of oil in 1991, respectively, with minimal exports. Turkmenistan's oil production grew to 12 million tons by 2021, but Uzbekistan's was essentially the same as in 1991 in 2014 to 2021 at about 2.8 million tons a year after a peak production level of 7.6 million tons in 1995.

TABLE 4.1 Central Eurasia: Annual Oil Production and Exports (thousand tonnes)

	Kazakhstan		*Turkmenistan*		*Uzbekistan*		*Azerbaijan*		
Year	*Production*	*Exports*	*Production*	*Exports*	*Production*	*Exports*	*Production*	*Exports*	
								Total	*To Europe*
1991	26,600	20,845	5,400		2,800				
1992	25,800	19,347	5,200		3,300				
1993	23,000	16,437	4,400		4,000	107			
1995	20,600	9,782	4,100		7,600	581			
2000	35,300	29,420	7,200		7,500	0			
2005	62,600	54,639	9,500		5,400	0			
2008	70,700	0	11,000		4,800				
2010	79,700	69,285	10,400		3,600	0			
2011	80,100		11,500		3,700		46.1		
2012	79,300	0	12,000		3,300		43.7		
2013	82,300		12,500		3,200		43.8		
2014	81,100	65,364	12,400		2,900	0	42.5		
2015	80,200	63,581	13,200		2,700	0	42.0		
2016	78,600	Na	13,200		2,600	na	41.4		
2017	86,900	Na	13,100		2,800	na	39.1		
2018	90,500		12,500		2,900		39.2		
2019	91,000		12,200		3,000		38.0		
2020	86,000		10,400		2,800		35.0		
2021	86,000		12,000		2,700		35.1	27.1	
2022							32.6	26.3	

Source: BP Statistical Review of World Energy, 2002-2022 (production) International Energy Agency (exports).

TABLE 4.2 Central Eurasia: Annual Natural Gas Production and Exports (billion cubic metres)

	Kazakhstan				Turkmenistan				Azerbaijan		
Year	*Production*	*Exports*			*Production*	*Exports*			*Production*	*Exports (by pipeline)*	
		Total	*To Russia*	*To China*		*Total*	*To Russia*	*To China*		*Total*	*To Europe*
2007	15.8				68.4	6.1					
2008	18.3				61.6	6.5					
2009	19.0	10.3	9.82	0	33.3	16.73					
2010	20.4	11.95	11.95	0	40.1	19.73					
2011	28.7	11.5	11.4	0	56.3	34.6			16.0		
2012	29.0	11.3	11.0	0	59.0	41.1			16.8		
2013	30.4	11.8	11.5	0.1	59.0	40.1		24.4	17.5		
2014	31.0	11.4	10.9	0.4	63.5	41.6		25.5	18.4		
2015	31.2	11.3	10.9	0.4	65.9	38.1		27.7	18.8		
2016	31.5	16.6	16.1	0.4	63.2	37.3		29.4	18.3		
2017	33.4	13.2	12.1	1.1	58.7	33.6		31.7	17.8		
2018	33.1	25.6	19.9	5.4	61.5	35.2		33.3	18.8		
2019	33.1	27.5	20.6	6.5	63.2	31.6		31.6	23.9		
2020	33.3	14.0	7.1	6.8	66.0	31.6	3.8	27.2	25.9		
2021	32.0	10.6	4.6	5.9	79.3	42.0	10.5	31.5	31.8	19.6	19.5

Source: BP Statistical Review of World Energy, 2002–2022.

Turkmen natural gas production was already very strong in 1991 at 78.6 billion cubic meters, while Kazakhstan produced 7.4 billion cubic meters. Production levels varied annually after that, but as indicated in Table 4.2. Kazakhstan's production nearly tripled to 20.4 billion cubic meters in 2010 and added at least another 10 billion cubic meters annually from 2014 to 2021. Turkmenistan's production was generally more than 60 billion cubic meters after 2007 and reached 79.3 billion cubic meters in 2021.

The natural gas export picture was more complicated; Russia, Iran, and China alternated as the main targets for exports at various points. As depicted in Table 4.2, Turkmenistan exported more than 5 billion cubic meters to Iran through much of the post-independence period until 2016; from 2009 until 2014, it also exported about 10 billion cubic meters to Russia. Turkmenistan then began substantial exports to China in 2010, rising to 31.7 billion cubic meters in 2017.

Kazakhstan's early natural gas exports largely went through the Russian network. In 2013, however, 100 million cubic meters were transported through the new pipelines to China. This grew to 1.1 billion cubic meters in 2017 and averaged more than 6 billion cubic meters annually from 2019 to 2021. Exports to Russia continued at a substantial level, above 10 billion cubic meters, and increased to 20.6 billion cubic meters in 2019, but then dropped dramatically in 2020 and 2021 to 7.1 and 4.6 billion cubic meters, respectively.

Uzbekistan's natural gas exports were divided between China and Russia beginning in 2014, after being mostly focused on the Russian network in the early years. In 2017, Uzbek natural gas exports to China amounted to 3.4 billion cubic meters, down from 4.3 billion cubic meters in 2016. Exports to Russia were 6.7 billion cubic meters, down from more than 10 billion cubic meters in 2009 and 2010.

Turkmenistan proceeded cautiously in its energy relationship with Russia initially while continuing a long-standing supply relationship with Iran. However, similar to Kazakhstan, it did establish a strong natural gas supply relationship with China beginning in 2010. What began as a lucrative supply relationship soured after a few years, given low global natural gas prices and the fact that the Chinese continued to deduct the costs for constructing the Turkmenistan-China pipeline from the natural gas import bill. The value of Turkmenistan's exports to China reportedly dropped from $10.4 billion in 2013 to less than $6 billion in 2016, while the volume of gas exported remained high. A key factor was the steep decline in gas prices in 2014. Natural gas exports to China from 2018 to 2021 were substantial, however, averaging more than 30 billion cubic meters per year (see Table 4.2).[34]

Turkmenistan has faced a price war with its neighbors in Kazakhstan and Uzbekistan for natural gas supplies to China and Russia. A 2017 decision by Russia's Gazprom to cancel its supply relationship with Turkmenistan apparently was based on much cheaper price quotes from the Kazakhs and Uzbeks; at about the same time, a contract dispute between Turkmenistan and Iran suspended that supply relationship as well. This left China as Turkmenistan's only customer. Plans to construct an additional gas pipeline from Turkmenistan to China (making a total of four in the network that connects all three supplier countries in Central Asia) have been shelved for the time being.[35]

With the decline in oil and gas prices after the peak of the early 2000s, Central Asian fossil fuel suppliers began to confront the realities of the resource curse. Diversifying their economies became a more serious preoccupation. This meant some attention to the support of new manufacturing enterprises, especially in Kazakhstan.[36] Subsequently, agricultural reform and modernization became an important focus, though not without important challenges and constraints.[37]

Pauline Jones Loung and Erika Weinthal's work *Oil Is Not a Curse* provides institutional context here by outlining the relationship between oil resources and institutions in Russia and the Central Eurasian economies. They argued that the extent to which oil wealth is managed properly has much to do with the character of its ownership. State ownership but private development of the oil wealth, creating a hybrid fiscal regime, in which areas closely connected to the foreign oil companies (FOCs), such as the transfer of revenues from the oil companies to the state, have been relatively transparent. In some cases, however, the taxation and expenditure regime for the rest of the mineral sector and economy remains opaque and unstable. Jones Luong and Weinthal argued that hybrid regimes are created by three important factors. First is the absence of transaction costs—in which the use of standard oil contracts provide the state with a constant stream of unrestricted revenue, creating few incentives to create broad-based taxation regimes. Second is high societal expectations that the oil wealth would be used to benefit the population at large, which leads to nontransparent and wasteful social spending. Finally, the avoidance of the obsolescing bargain—the tendency for host countries to change the terms of oil contracts in their favor once the foreign oil companies have installed their fixed capital and have done the difficult technical work. Commonly, the presence of a dominant firm to represent private multinational stakeholders allows the FOCs to pressure the government to enact fiscal reforms.[38]

The challenge of answering the question of why resource-rich countries often have failed to develop economically remains perplexing. Sachs and Warner produced the first groundbreaking study on what became known as the resource curse, arguing that "economies with a high ratio of natural resource exports to GDP in 1971 (the base year) tended to have low growth rates during the subsequent period 1971–89."[39]

The main mechanism for the curse was the Dutch disease—named for the economic woes that plagued the Netherlands after it discovered and exploited huge reserves of natural gas in the 1960s and 1970s. The "disease" occurs when investment in the natural resources sector crowds out investment in the tradable manufacturing sector, leading to the decline of the latter sector and the rise of the non-tradable goods sector (often subsistence agriculture).[40] While the aforementioned situation is not inherently pernicious, economic development slows if the manufacturing sector has unique strong forward and backward linkages to the economy and a tendency to promote learning-by-doing positive externalities—both of which are not replicated in the natural resource or non-tradable goods sectors.[41]

Lederman and Maloney in a 2007 World Bank publication argued that "overall, natural resource wealth is good for development. However, more important… natural resource wealth becomes a real development asset when coupled with investments in skills and technological capabilities with good macroeconomic institutions and management."[42] Specifically, they argued that much of the low growth witnessed by resource-rich countries was "deficient technological adoption" caused by "low investment in human capital and scientific infrastructure" and "inward-looking industrialization" that "discouraged innovation and created a sector whose growth depended on artificial monopoly rents" and "undermined natural resource-intensive sectors that had the potential for dynamic growth."[43] They also noted the importance of "spillovers or linkages across firms that produce increasing returns to scale," domestic infrastructure, the quality of public institutions and governance, investments in knowledge accumulation…macroeconomic volatility…and trade intensity."[44]

Other theorists of economic development expounding upon the resource curse include Richard Auty. Auty delineated a "staple trap model," in which high oil rents cause several pernicious effects. First, high oil rents incentivize the state to increase its revenue through capturing rents rather than wealth creation, increasing the rewards for capturing the state (and its rent stream). Second, high oil rents cause the economy to become dependent on the primary sector for a much longer period of time, miss out on the benefits of labor-intensive manufacturing, and face a high surplus of labor that must be co-opted through overextension of the bureaucracy and protection of infant industries, which themselves further hinder economic development. Third, high oil rents cause weakened sanctions against anti-social governance, as the growth of social entitlements, rent-driven urbanization, and business rent seeking create constituencies dependent on the government and therefore unwilling to provide pressure for accountability.[45]

The developmental state is a term most commonly used to describe the state as a key driver of economic development in Japan and later, the newly emerging economies (NIEs) of South Korea, Taiwan, Hong Kong, and Singapore. Chalmers Johnson argued that the developmental state consisted of the unique and intimate relationship between government bureaucracy and industry that resulted in a "plan-oriented market economy."[46]

The relatively state-centric process of development in Kazakhstan suggests that a careful examination of the archetype of successful state-facilitated development is in order. *The Myth of the Powerless State* by Linda Weiss does just this. Reviewing the role of the state in the economic development of Germany and Japan, Weiss crystallized the definition of state capacity as "the ability of policy-making authorities to pursue domestic adjustment strategies that, in cooperation with organized economic groups, upgrade or transform the industrial economy." This could emerge in the form of structural shifts from "declining to expanding sectors" or "technological diffusion and innovation" as well as the "creation of new industries, products, and processes."[47] Weiss then argued the pinnacle of state-led development in the modern era is governed interdependence: "a negotiated relationship, in which public and private participants maintain their autonomy, yet which is nevertheless governed by broader goals set and monitored by the state."[48]

Finally, Mariana Mazzucato extended the model of the developmental state even further. She argued that the ideal type of a state is to be entrepreneurial—funding risky, long-term technical projects and thinking big: attempting to create markets for what does not currently exist. While her chief example of the power of the entrepreneurial state—extensive US government funding for basic research that developed many of the technologies available in the iPhone and other important communication and computer devices—comes from the United States,[49] her model can be extended to other countries, especially Kazakhstan and Uzbekistan. As the entity in each case with the greatest resources and the mandate to provide public goods and upgrade the economy, the state can afford financially to be entrepreneurial. Further, although subject to many other restraints, the relative dearth of deep democratic accountability and oversight would lend themselves to greater government experimentation and room to fail.

A comprehensive assessment of the extent to which these factors are at work in fossil fuel rich economies of Central Asia is beyond the scope of this chapter, but some should be kept in mind as we strive to understand the challenge of economic diversification, including recent attention to agricultural development and modernization. The Kazakh and Uzbek cases are particularly interesting in this respect, as we focus on the water-energy-food nexus.

Addressing the Cotton Monoculture and Modernizing Agricultural Development

The picture that is commonly drawn about the water resource constraints facing Central Asia is to cast primary blame on the apparent (irrational?) long-term commitment Kazakhstan, Tajikistan, Turkmenistan, and especially Uzbekistan have to cotton cultivation as a/the primary focus of agricultural production. Notwithstanding increased analysis of the impact of climate change, especially real and anticipated glacier melt and precipitation change, the policy discussions often emphasized alternative cropping and dramatic improvements in irrigation efficiency. In the decade or so following independence, concrete progress in these areas was modest. Writing in 2005, the International Crisis Group argued somewhat alarmingly:

> The environmental costs of the monoculture have been devastating. The depletion of the Aral Sea is the result of intensive irrigation to fuel cotton production. The region around the sea has appalling public health and ecological problems. Even further upstream, increased salinization and desertification of land have a major impact on the environment. Disputes over water usage cause tension among Central Asian states....The cotton monoculture is more destructive to

> Central Asia's future than the tons of heroin that regularly transit the region. Although the international community has invested millions of dollars in counter-narcotics programs very little has been done to counteract the negative impact of the cotton industry. Changing the business of Central Asian cotton will take time, but a real reform of this sector of the economy would provide more hope for the stability of this strategic region than almost anything else the international community could offer.[50]

Table 4.3 depicts the record of cotton production and exports in Central Asia since independence. Uzbekistan has led the region in both categories by far, ranked in the top ten of producers and exporters globally each year. Turkmenistan is next, followed by Tajikistan and Kazakhstan. However, all four have seen a substantial decline in cotton production since independence. Kyrgyzstan produces cotton as well, but the totals have not been among the top 30 globally in any year from 1989 to 2022.

The decline in Central Asian cotton production is promising; the Soviet legacy appears to be weakening, and the governments seem to be motivated to diversify their agriculture cropping, especially in recent years. For Uzbekistan and Kazakhstan, the change is in part a consequence of explicit policy and program reforms. Uzbekistan seeks to make its agriculture more sustainable and less dependent on water and manual labor, and there is a focus on producing smaller amounts of high-quality cotton in order to support the export of finished high value textiles and clothing, instead of raw cotton. Kazakhstan seeks more diverse and productive agriculture for rural development and to help adjust to low or at least volatile global prices of oil and gas. There is new energy in the agriculture development effort, after many years of relative neglect. Former President Nursultan Nazarbayev began to stress this diversification priority in his annual messages on the economy, most notably in 2016 and 2017. See also his December 15, 2012 message: *Kazakhstan 2050 Strategy*.

These reforms include efforts to rebuild the Aral Sea fishery and to explore the prospect for developing a sustainable meat export industry by finding ways to use the vast Northern grasslands for expanded cattle grazing.[51] President Tokayev has also raised the prospect for a more independent and "multi-vectored" trade policy by criticizing the Eurasian Economic Community at a session of the Eurasian Economic Forum in Moscow. He warned against efforts to politically integrate the EaEU membership. He stated that Kazakhstan sees the union as an exclusively economic bloc. "This is precisely what it says in the 2015 treaty."[52]

Turkmenistan, which of course is not a member of the EaEU, has made similar noises that the fossil fuel price volatility required a substantial campaign of economic diversification, including targets for more diverse agricultural production. As yet, however, there have been no detailed strategy documents released publicly.

The Uzbek Ministry of Agriculture noted that while Uzbekistan has for some time been at least the sixth largest producer and fifth largest exporter of cotton in

TABLE 4.3 Central Asia: Annual Cotton Production and Exports

	Central Asia: Annual Cotton Production and Exports												
	(000) 480-pound bales												
	1989	*1990*	*1991*	*1992*	*1995*	*2000*	*2005*	*2010*	*2015*	*2016*	*2017*	*2018*	*2022–2023*
Uzbekistan													
Production	7,605	7,317	6,628	5,851	5,740	4,400	5,550	4,200	3,800	3,725	3,850	3,750	3,100
Exports	6,810	5,393	5,200	5,500	4,524	3,450	4,800	2,650	2,200	1,500	1,100	1,000	
Turkmenistan													
Production	1,823	2,007	1,971	1,791	1,150	825	975	1,750	1,450	1,325	1,360	1,400	900
Exports	1,781	1,743	1,675	1,850	919	675	550	725	1,250	850	700	625	
Tajikistan													
Production	1,332	1,176	1,134	57	550	465	634	409	413	430	550	530	500
Exports	1,118	824	1,000	600	450	407	592	354	450	300	500	450	
Kazakhstan													
Production	455	468	432	350	340	400	635	275	200	300	290	300	290
Exports	315	285	300	250	245	400	829	265	185	285	280	200	

Source: National USDA & National Cotton Council.

the world, there has been a long-term effort to steadily reduce the area devoted to cotton (from 1.83 to 1.071 million hectares 1991 to 2018); the 2016–2020 plan was to reduce the area of cotton cultivation by 170,000 hectares and convert it to production of grains, vegetables, potatoes, oilseeds, and fruits. By 2021, the area devoted to cotton was targeted to be .95–1.0 million hectares. Cotton production did not decline dramatically, because the Ministry encouraged much more intensive and mechanized cultivation. In the early years after independence, only 4–5% of the cotton produced was processed in Uzbekistan; now that figure is 40–50%. Since at least 2016, the main focus in Uzbekistan has been on processing the raw cotton produced in modern gin mills. Figure 4.1 depicts the significant increase in fruit production in Uzbekistan, as cultivation priorities changed. Orchards increased from 138,600 to 272,700 hectares in the same period, and gross volume of fruit production increased from 516,800 to 3,076,300 tons.[53]

The privatization and expansion of non-cotton farming in Uzbekistan has not been without challenges; new and especially young farmers needed training, and there was a strong motivation to reduce chemical pesticide and herbicide use. Organic farming and integrated pest management have been introduced, but training and research into new methods need attention. The Ministry of Agriculture is working closely with key Uzbek agriculture universities since the election of President Shavkat Mirziyoyev on reforms and agriculture development. The principal institutions are Tashkent State Agricultural University with its branches in Andijan, Nakus, and Termez; the Tashkent Institute of Irrigation and Agricultural Mechanization Engineers, with its branch in Bukhara; and the Samarkand Veterinary Medical Institute. There seemed to be real commitment to development of an elaborate network of some 41 vocational "colleges" which would function as an effective extension system, much like land grant universities have established in most states in the US. The plan was to dedicate 13 for agriculture, 13 for water management, 13 for veterinary and livestock production support, and 2 for fisheries.

A key element of the Uzbek agricultural reform effort was the commitment in recent years to reduce child labor in the harvesting of cotton. Jonas Astrup of the UN's International Labour Organization has argued that the implementation of labor reforms in Uzbekistan under President Mirziyoyev must be regarded as a clear success story. According to an ILO assessment: "Almost two million people are recruited every year for the annual cotton harvest in Uzbekistan. The country has succeeded in eradicating systemic forced labor and systemic child labor during the 2021 cotton production cycle...."[54]

Capture fisheries have not really revived in Uzbekistan, indeed in Central Eurasia generally, since the dissolution of the Soviet Union, but aquaculture operations have expanded and modernized. In Uzbekistan, 307,000 hectares were devoted to aquaculture, producing some 80,000 tons of fish in 2017. The Ministry's projection was that 150,000 tons would be produced by 2021. 282,000 hectares of aquaculture are managed in natural reservoirs; 25,000 hectares are operated in artificial ponds.[55]

Uzbekistan's greatest challenge, along with most of Central Eurasia, was perhaps to overcome the Soviet legacy in agriculture and economic backwardness. For Uzbekistan, this was in large part the Soviet (and Tsarist) assignment to the role of main cotton producer. The Mirziyoyev reforms have offered some hope to maximize the promise of substantial and more diverse agricultural productivity. There remain substantial domestic economic interests at play, but the new regional opportunities, given Russia's alarming actions in Ukraine and President Mirzivoyev's apparent commitment to change, provide reason for optimism.[56]

The 2016–2020 plan envisioned further reductions in the land under cotton cultivation, replacing the fiber with edible crops such as vegetables, potatoes, oilseeds, fodder, and other crops, including orchards and vineyards. One can point to some important progress in diversifying agricultural cultivation and production in the face of water constraints. For example, in 1991, the area under cultivation for grains in Uzbekistan was 870,000 hectares with an average yield of .6 to .7 tons per hectare. At that time, more than 5 million tons of grain and grain products were imported annually to meet the needs of the population. A combination of efforts to increase yields and expansion of the are cultivated led to dramatic gains in production and this less dependence on exports and reduced reliance on the cotton monoculture. By 2017, the volume of grain produced increased to 8.1 million tons with an average yield of 5.5 tons per hectare in irrigated areas.[57]

Vegetable production in 2017 amounted to 11.4 million tons (3.4 times more than in 1991). Since independence, more than 70 new varieties of vegetables, 20 varieties of melons, and 10 varieties of potatoes were created and introduced. Similarly, there has been substantial growth in the production of fruit crops. Figure 4.1

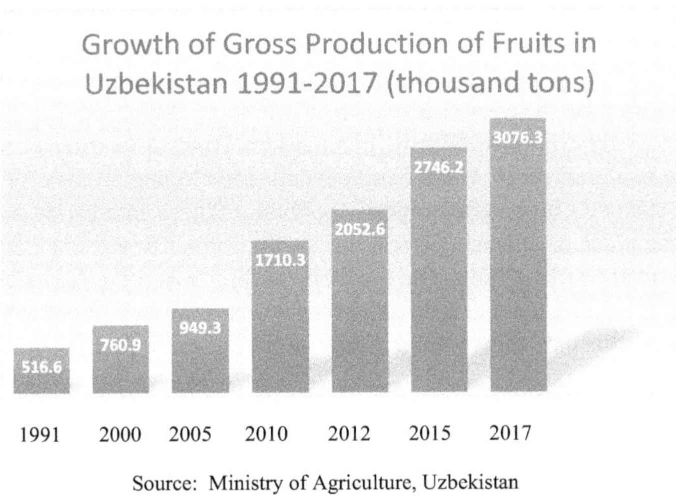

FIGURE 4.1 The Growth of Gross Production of Fruits in Uzbekistan 1991–2017.

Source: Ministry of Agriculture, Uzbekistan.

depicts the dramatic rise from only 760,900 tons in 1991 to more than 3 million tons in 2017. Considerable attention has been paid in recent years to sustainable irrigation techniques in this effort. Yields in orchards have increased to 6.16 tons per hectare while vineyards reached 6.21 tons per hectare due to incentives introduced promoting their expansion, timely implementation of best agri-technical practices and planting of high-quality fruit and vine seedlings over the last ten years.[58]

The torturous history of developing a more diversified economy and achieving general food security has faced numerous challenges and missteps. For example, the ups and downs of the Kazakh fisheries industry are detailed in a recent article by a joint team of researchers from Al-Farabi Kazakh National University and Michigan State University.[59]

President Kassym-Jomart Tokayev has provided new attention to the challenge of economic diversification. Shortly after being designated as Nazarbayev's successor, he promised to bring new accountability to governance in Kazakhstan, but this was tested severely during the violent domestic energy price protests that erupted in January of 2022. His orders to "shoot to kill" without warning, given his argument that foreign terrorists were fomenting the violence, gave rise to considerable external and domestic criticism. A range of reforms followed the stability that Russian "peace keeping" support helped to provide under the CSTO ties. Tokayev won re-election handily with 81.3% of the vote in a snap election held November 20, 2022, and has continued both his somewhat independent foreign policy and domestic reform agenda. These reforms have included a strong agricultural development orientation.

On the positive side, there appears to be real promise in entrepreneurial efforts to modernize aquaculture. For example, a sturgeon farm was established in 2022 in the village of Koldi, about 90 kilometers from Almaty. The design and support of the system was modern and well-staffed with a plan to help recover significant production of sturgeon and beluga caviar by 2025. It includes an exterior fish pond which may be the site of development for sport fishing outings for tourists and business retreat groups. Nearby, is a modern Trout farm, already in production for the fish markets of Almaty and beyond. Developing an export market for China, Russia, and Europe is the objective of both of these enterprises. Essentially, they take the place or at least compensate for the former Soviet state aquaculture establishment at nearby Chilik. The Chilik pond farm was privatized after the collapse of the Soviet Union, but it did not prosper under the lack of government support for modernization and commercialization in the early years of Kazakhstan's transition.[60]

It is often argued that the key obstacle poorer Central Eurasian states face is the lack of investment funds to modernize and diversify agriculture and aquaculture. Indeed, there is a tendency to focus the limited financial and manpower resources on crops like rice, tobacco, and wheat in Kyrgyzstan.[61]

As stressed before, this is complicated for countries that depend so much on trans-border water sources. Kazakh officials and environmental researchers commonly note that 80% of Kazakhstan's water comes from foreign sources. In addition to the Aral Sea crisis discussed earlier, one should note the rising threat to the sustainability of the Ili River/Lake Balkhash Ecosystem. The water of the Ili River and its tributaries begins its flow high in the Tian Shan mountains of Xinjiang, China, making its way through an increasingly developed Chinese agricultural area before reaching the Kazakh border near Yining.[62] There is some evidence that the water flow from China has been reduced and that the dynamics of the fish population in Lake Balkhash have changed substantially. It is a complicated ecosystem affected by glacier melt in the Chinese mountains as well as a Soviet hydroelectric system built in 1969 in what is now Kazakhstan creating the Kapchagay Reservoir. The future of Lake Balkhash may not be as dire as what led to the desiccation of the Aral Sea, but it is clear that equitable trans-border water sharing remains elusive. A similar challenge is evident in the Irtysh River ecosystem which involves concerns about increased Chinese consumption before the water reaches Kazakhstan and Russia. Some effort to engage the Chinese and Russians in a tripartite negotiation on equitable sharing has thus far had little agreement.[63]

Chinese-Kazakh negotiations on water sharing arrangements have been underway for more than fifteen years with little concrete result. Some have argued that while the reduced water flow via the Ili River to Lake Balkhash has had an important impact on the associated Kazakh fisheries and agriculture, the Kazakh government has not pressed China for a proper sharing agreement vehemently, given the importance of other geopolitical considerations at present. On the other hand, the centrality of this region of Kazakhstan to the success of China's Belt and Road Initiative (BRI) for a new Silk Road to Europe would seem to provide the Kazakhs with important bargaining leverage.[64]

Tajikistan offers the clearest example of urgent hydroelectric generation capacity development, sometimes without due consideration to rural population and fish habitat impacts. A large number of construction and rehabilitation projects have been initiated. For Tajikistan, the key dam constructions were the Rogun and Sangtuda hydropower projects. Rogun was begun during the Soviet period but abandoned with independence and the turbulent civil war. Resurrected, it is set to solve Tajikistan's annual winter energy crisis with an expected installed capacity of 3,600 Megawatts. Unfortunately, sited on the Vakhsh river, a key tributary of the Amu Darya, it threatened to displace 42,000 people from surrounding mountain villages. The World Bank was slow in approving the project,[65] but President Mirziyoyev softened Uzbek opposition to the project.[66] Uzbek researchers and officials see little prospect, however, for the government to embrace a system of direct compensation for water supplies from rivers sourced in Tajikistan. The notion of trading energy supply for water supply as a direct commercial deal seems off the table for now.

Prosperity and Environmental Sustainability in Central Eurasia

The World Bank's Europe and Central Asia Economic Update, *Weak Growth, High Inflation, and a Cost-of Living Crisis* for Spring 2023 did not offer a rosy picture for Central Asia and the South Caucasus

> The Russian Federation's invasion of Ukraine has had devastating economic and social consequences for Ukraine and profound impacts on development and poverty alleviation in Europe and Central Asia (ECA). Growth in economic activity in the emerging markets and developing economies of ECA slowed substantially in 2022 to 1.2 percent, as surging inflation, energy and value chain disruptions, and marked monetary tightening weighed on the economic expansion....
>
> Rising food prices and energy bills, higher mortgages and rents, and elevated prices for services are eroding the standards of living for a large part of the population of ECA.[67]

The ECA region summarized here includes Russia, Ukraine, Belarus, and Moldova all of which have been affected by the invasion very negatively, along with Central Europe, where there is clear stagnation. Central Asia and the South Caucasus (what we together call Central Eurasia) have seen some reasons for optimism in their 2022 economic performance and near-term prospects. Table 4.4 indicates economic growth in 2022 was quite strong in seven of the eight Eurasian countries (Turkmenistan remains a black box in many respects). Kazakhstan was at the head of the pack in Gross Domestic Product per capita in 2021 with $10,374. Azerbaijan, Georgia, and Turkmenistan were all above $5,000. Tajikistan, Kyrgyzstan, and Uzbekistan measured less than $2,000 per capita. Poverty rates in Central Eurasia, remarkably, were down somewhat, despite the disruption and conflict in the larger region. Azerbaijan's rate was not available to the World Bank. Russia's rose slightly in 2022. Perceived corruption in the region, according to Transparency International, suggested some comparative progress in Georgia, Armenia, Kazakhstan, and Uzbekistan, while the others ranked among the worst in the world.

Conclusion

Certainly, there is no common willingness at present to accept supranational authority and effective regional regulation to respond to the water-energy-food challenges in Central Eurasia. While individual states enjoy substantial income from oil and natural gas exports at times, the prospect for a substantial regional pool of financial resources for investment in environmental remediation and sustainability projects seems unlikely in the near term. Scientific expertise and technical assistance possibilities are available, but national financial resources and political will have been in short supply. Fortunately, the region has drawn the attention of

TABLE 4.4 Central Eurasia: Economic Development Indicators

Country	*Economic Growth*	*GDP/Capita*	*Poverty %*[a]		*Perceived Corruption*	
	(2022)	*(2021)*	*(2021)*	*(2022)*	*TI Index*	*Rank*
Armenia	12.6%	$4,967	51.7%	41.8%	46/100	63/180
Azerbaijan	4.6%	$5,388	6.0[b]		23/100	157/180
Georgia	10.1%	$5,023	55.4	52.2	56/100	41/180
Kazakhstan	3.2%	$10,374	16.7	15.8	36/100	101/180
Kyrgyzstan	7.0%	$1,277	62.2	59.6	27/100	140/180
Tajikistan	8.0%	$897	50.6	46.8	24/100	150/180
Turkmenistan	na	$7,345	Na	na	19/100	167/180
Uzbekistan	5.7%	$1,983	17.0	14.0	31/100	126/180
Russian Federation	−2.1%	$12,195	3.1	3.2	28/100	137/180

[a]At US $6.85 in 2017 ppp (upper middle-income).
[b]From State Statistical Committee of Azerbaijan; not reviewed by the World Bank.
Sources: World Bank ECA, Spring 2023 and Transparency International Survey.

external donors like the Asian Development Bank and the World Bank, which seem committed to a regional approach to the water crisis; China's Asian Infrastructure Investment Bank and its BRI strategy may also be an important source of support, if the commercial motivations can be harnessed to an ecologically sound development strategy.

The way forward for Central Eurasia's energy and water challenges is relatively straightforward, albeit politically challenging: (1) adopt and enforce regional and complementary national rules to prevent, excessive irrigation, pollution, and overfishing; (2) address trans-border water sharing, conservation, and management aggressively on a regional basis before the resources are degraded beyond recovery—the partial recovery of the Kazakh portion of the Aral Sea is a promising start; (3) continue to expand aquaculture research, development, and commercialization to replace collapsed fisheries that cannot be revived; and (4) expand essential hydroelectric generation capacity in Kyrgyzstan and Tajikistan, but do so within a framework of environmental impact assessment that includes consideration of alternative strategies to reduce potential fish habitat loss, as well as efforts to limit human dislocation and trans-border tensions. The way is straightforward, but the required level of political commitment and compromise will not come easy, given the nature of the Central Eurasian regimes.

Martha Olcott argued in 1996 that a confederation of Central Asian states was a desirable way forward to meet some of the challenges the emerging independent states faced. Closer cooperation certainly would be desirable to address the common environmental and external challenges apparent now—more than 25 years after she made that suggestion. Olcott noted wisely that such an organization would not be feasible in the face of Russian opposition, and that the Russians were adept at manipulating the regional rivalry evident between Presidents Nazarbayev and Karimov—a second constraint on regional cooperation. She went on to argue that Turkmenistan was the true regional spoiler to productive cooperation.[68] Now that Russia is perhaps distracted in its position in the region, especially vis-à-vis China, and Karimov and Nazarbayev have been succeeded by the less combative and more reform-minded Presidents Mirziyoyev and Tokayev, one might argue that a confederation is more viable. Turkmenistan also may not be such a spoiler now, given the character of its leadership, after the passing of Turkmenbashi, and the economic constraints that seem to be growing even more severe in the environment of low global natural gas prices.

Notes

1 Sven Beckert, *Empire of Cotton: A Global History* (New York: Vintage Books, 2014), pp. 376–378.
2 See: Sven Beckert, "Emancipation and Empire: Reconstructing the Worldwide Web of Cotton Production in the Age of the American Civil War," *The American Historical Review* 109, no. 5 (December 2004): pp. 1405–1438; and *Empire of Cotton*, 2014; Igor Lipovsky, "The Central Asian Cotton Epic," *Central Asian Survey* 14, no. 4 (1995); and L. Pavlov, "Cotton in Russia," *The Russian Review*, Vol I, May, 1916.
3 Martin McCauley, *Khrushchev and the Development of Soviet Agriculture: The Virgin Land Programme 1953–1964* (New York: Holmes & Meier Publishers, Inc.1976), chapters 1 and 2. See also: Donald W. Treadgold, *The Great Siberian Migration: Government and Peasant in Resettlement from Emancipation to the First World War* (Princeton, NJ: Princeton Legacy Library, 1957).
4 McCauley, p. 25.
5 Zhores A. Medvedev, *Soviet Agriculture* (New York: W.W. Norton & Co.1987), p. 60; see also Ch. 3.
6 Ibid., p. 96.
7 Lewis Siegelbaum, "Great Fergana Canal." Seventeen Moments in Soviet History. http://soviethistory.msu.edu/1939-2/great-fergana-canal/.
8 McCauley, p. 31.
9 Pannier, Bruce and Edige Magauin. "Kazakhstan: China Discusses Future of Irtysh River." RadioFreeEurope/RadioLiberty Archive, May 9, 1999. https://www.rferl,.org/a/1091429.html.
10 Medvedev, pp. 411–412.
11 McCauley, p. 54.
12 Peter R. Craumer, "Trends in Soviet Dryland Farming and Soil Conservation Practices with Comparison to North American Developments," in *Soviet Agriculture: Comparative Perspectives*, ed. Kenneth R. Gray (Ames: Iowa State University Press, 1990), p. 176; see also McCauley, Ch. 4–6.
13 William Taubman, *Khrushchev: The Man and His Era* (New York: W.W. Norton, 2004), pp. 262–269 and 605–609.
14 Martha Brill Olcott, *The Kazakhs* (Stanford, CA: Stanford University, Hoover Institution Press, 1995), Second edition, Ch. 10.
15 Stefan Hedlund, *Private Agriculture in the Soviet Union* (London: Routledge, 1989), especially Chs. 1, 6 and 7. For more general discussions of these challenges, see: Norman A. Graham, Folke Lindahl, and Timur Kocaoglu, *Making Russia and Turkey Great Again? Putin and Erdogan in Search of Lost Empires and Autocratic Power* (Lanham, MD: Lexington Books, 2021), pp. 31–35. Expansive and perceptive analyses of these

challenges (and the failed Soviet and Russian response) can be found readily in the multiple volumes authored by Anders Aslund (2002, 2007, and 2019) and Marshall Goldman (1968, 1983, 1991, 1996, 2003, and 2008).

16 D. Gale Johnson and Karen McConnell Brooks, *Prospects for Soviet Agriculture in the 1980s* (Bloomington: Indiana University Press, 1983 (in Association with the Center for Strategic and International Studies, Georgetown University)), pp. 73–91 and 195–204.

17 Karen M. Brooks, "Soviet Agricultural Policy and Pricing under Gorbachev," in *Soviet Agriculture: Comparative Perspectives*, ed. Kenneth R. Gray, pp. 127–129; see also Litvin, Valentin. *The Soviet Agro-Industrial Complex: Structure and Performance* (Boulder, CO: Westview Press, 1987), pp. 141–148.

18 Johnson and Brooks, pp. 198–204.

19 See, for example: Aslund (2002, 2007, 2019), Lavigne (1999); Olcott (2005, 2010, 2012), Overland et al. (2010), Spechler, Ahrens, and Hoen (2017), and Peimani (2021).

20 Siegmar-W. Breckle, Walter Wucherer, Liliya A. Dimeyeva, and Nathalia P. Ogar (eds.), Aralkum-*A Man-Made Desert: The Desiccated Floor of the Aral Sea* (Heidelberg: Springer, 2012). See also: Philip Micklin, N.V. Aladin, and Igor Plotnikov (eds.), *The Aral Sea: The Devastation and Partial Rehabilitation of a Great Lake* (Berlin: Springer, 2014).

21 Islam Karimov, *Uzbekistan on the Threshold of the Twenty-First Century: Challenges to Stability and Progress* (New York: St. Martin's Press, 1998), pp. 71–81.

22 FAO-FID/CP/UZB, Rev.1 2003, and FAO-FAC/1020/1 2009; see also Graham (2015) and Graham, Pueppke and Uderbayev (2017) for assessments of recent trends in Central Asian fisheries.

23 World Bank, *Kazakhstan: Transition of the State*, 1997, pp. ix–26; see also Olivier Roy, *The New Central Asia: The Creation of Nations* (New York: New York University Press, 2000).

24 International Crisis Group, 2002, pp. i–ii. See also: ICG Asia Report No. 93 2005.

25 World Bank, Europe and Central Asia Region. *Water Energy Nexus in Central Asia: Improving Regional Cooperation in the Syr Darya Basin* (Washington, 2004), pp. 4–11.

26 Casey Michel. "Tajikistan's Rogun Dam Rankles Uzbekistan." *The Diplomat*, July 21. 2016; see also: ICG, 2002, p. 23.

27 International Crisis Group. *The Curse of Cotton: Central Asia's Destructive Monoculture*. Asia Report No. 93. Bishkek/Brussels: February 28, 2005.

28 See: ICG, 2002, pp. 24–29; and World Bank, 2004, pp. 12–22.

29 Andrea Zinzani. *The Logics of Water Policies in Central Asia: The IWRM Implementation in Uzbekistan and Kazakhstan* (Zurich, Switzerland: LIT Verlag, 2014), p. 224.

30 Ariel Cohen. *Kazakhstan: The Road to Independence—Energy Policy and the Birth of a Nation* (Washington, DC: Johns Hopkins SAIS/Silk Road Studies Program, 2008), p. 130.

31 Steve Levine, *The Oil and the Glory: The Pursuit of Empire and Fortune on the Caspian Sea* (New York: Random House, 2007), pp. 380–385.

32 Ibid., p. 383.

33 Aygerim Sarymbetova, "Kazakhstan to Start Gas Deliveries to China." *Caspian News*, October 10, 2017.

34 Catherine Putz, "China's Imports of Turkmen Gas Decline." *The Diplomat*, December 27, 2017.

35 Bruce Pannier, "Turmenistan's Gas Gloom." *OilPrice.com*, March 11, 2017.

36 Gauhar Abdygaliyeva, et al. *Economic Diversification in the Republic of Kazakhstan Through Small and Medium Enterprise Development: Introducing New Models of Finding for SMEs* (New York: Columbia University, School of International and Public Affairs, 2007).

37 See the detailed assessments of the general challenges for Central Asia by Olcott, 2005; and of Kazakhstan specifically in Olcott, 2010.

38 Pauline Jones Luong and Erika Weinthal, *Oil Is Not a Curse* (New York: Cambridge University Press, 2010), pp. 193–194, 201, and 212.
39 Sachs, J., and A. Warner, "Natural Resource Abundance and Economic Growth." NBER Working Paper Series (Working Paper 5398), 1995, p. 1.
40 Ibid, p. 6.
41 Ibid.
42 Lederman, D., and W. Maloney, *Natural Resources, Neither Curse Nor Destiny* (Palo Alto, CA: Stanford Economics and Finance, an imprint of Stanford University Press, 2007), p. xiv.
43 Ibid., p. 7.
44 Ibid., p. 9.
45 Auty, R. "Resource-Driven Models of the Development of the Political Economy," in *Energy, Wealth and Governance in the Caucasus and Central Asia: Lessons Not Learned*, eds. R. Auty and I. de Soysa (New York: Routledge, 2006), p. 26; see also Chapter 15 for a pessimistic assessment of what lessons have not been learned on this.
46 Chalmers Johnson, *MITI and the Japanese Miracle* (Stanford, CA: Stanford University Press, 1982), p. 1.
47 Linda Weiss, *The Myth of the Powerless State* (Ithaca, NY: Cornell University Press, 1998), p. 5.
48 Ibid., pp. 5 and 38.
49 Mariana Mazzucato, *The Entrepreneurial State* (New York: Anthem, 2013).
50 International Crisis Group, 2005, pp. i–ii.
51 Steven G. Pueppke. "Kazakhstan's Northern Steppe Grasslands: An Underutilized Resource to Increase Livestock Production," in *The New Kazakhstan: Steering the Steppe Amid a Global Storm*, ed. Marc Uzan (Paris: Reinventing Bretton Woods Committee, 2023).
52 Kassym-Jomart Tokayev. Public statement quoted in *The Moscow Times*, May 25, 2023.
53 Data presented in briefing by Ministry of Agriculture officials, *Tashkent*, May 23, 2018.
54 Jonas Astrup to CNN, May 26, 2023. See also: "Uzbek Cotton Is Free from Systemic Child Labour and Forced Labour." *ILO News*, 1 March 2022. https://www.ilo.org/global/about-the-ilo/newsroom/news/WCMS_838396/lang--eng/index.htm
55 See: Norman A. Graham, Steven G. Pueppke, and Talgat Uderbayev, "The Current Status and Future of Central Asia's Fish and Fisheries: Confronting a Wicked Problem," *Water* (August 2017); and Norman A. Graham, Stephen Pueppke, and Sabyr Nurtazin, et al., "The Changing Dynamics of Kazakhstan's Fisheries Sector: From the Early Soviet Era to the Twenty-first Century," *Water* 14 (2022).
56 Mamuka Tsereteli, "The Economic Modernization of Uzbekistan," in *Uzbekistan's New Face*, ed. S. Frederick Starr and Svante E. Cornell (Lanham, MD: Rowman & Littlefield, 2018), pp. 98–107.
57 Uzbekistan, Ministry of Agriculture. *Progress Report on the Last Quarter Century*, May, 2019, p. 8.
58 Ibid., pp. 9–11.
59 Norman A. Graham, Stephen G. Pueppke, Sabir Nurtazin, Talgarbay Konysbayev, Farid Gibadulin, and Meiirli Sailauov, "The Changing Dynamics of Kazakhstan's Fisheries Sector: From the Early Soviet Era to the Twenty-first Century," *Water* (May, 2022).
60 Visit of installations and interviews with Professor Sabyr Nurtazin and Fish biologist and investor Talgarbay Konysbaev, June 3, 2023.
61 Food and Agriculture Organization. FAO-FID/CP/KGZ 2007:2.
62 Stephen G. Pueppke, Sabir T. Nurtazin, Norman A. Graham, and Jiaguo Qi. "Central Asia's Ili River Ecosystem as a Wicked Problem: Unraveling Complex Interrelationships at the Nexus of Water, Energy, and Food," *Water* (April, 2018). See also: Norman A. Graham, Stephen Pueppke, and Talgat Uderbayev. "The Current Status and Future of Central Asia's Fish and Fisheries: Confronting a Wicked Problem," *Water* (August 2017).

63 Albina Zhanbosinova, Fatima Kukeyeva, Tolganay Ormysheva, and Norman A. Graham. "Irtysh Whirls: The Geopolitics of River Rights Between Kazakhstan, China and Russia." CERES Occasional Papers. Center for European, Russian and Eurasian Studies. Michigan State University, July, 2019.
64 See: Fatima Kukeeva and Tolganay Ormysheva. "The Ili River: Is It a 'Casualty' of Kazakhstan's Complicated Relationship with China." Paper presented at the Annual Conference of the Central Eurasian Studies Society, University of Washington, Seattle, October, 2017, for an interesting discussion of the history of the negotiations.
65 World Bank, *Assessment Studies for Proposed Rogun Hydropower Project in Tajikistan*, Washington, DC, September 1, 2014.
66 Catherine Putz, "Uzbekistan's Changing Rogun Tone." *The Diplomat*, July 10, 2017.
67 World Bank, *Weak Growth, High Inflation, and a Cost-of-Living Crisis: Europe and Central Asia Economic Update* (Office of the Chief Economist, Spring 2023), pp. ix–x.
68 Martha Olcott. *Central Asia's New States: Independence, Foreign Policy and Regional Security* (Washington, DC: U.S. Institute of Peace Press, 1996), pp. 144–152.

References

Abdygaliyeva, Gauhar, et al. *Economic Diversification in the Republic of Kazakhstan through Small and Medium Enterprise Development: Introducing New Models of Finding for SMEs*. New York: Columbia University, School of International and Public Affairs, 2007.

Acemoglu, Daron, S. Johnson, and J. Robinson. "Institutions as a Fundamental Cause of Long-Run Growth," in *Handbook of Economic Growth*, edited by P. Aghion and S. Durlauf (1st ed.). Amsterdam: North Holland, 2005, pp. 385–472.

Acemoglu, Daron, and James A. Robinson. *Why Nations Fail: The Origins of Power, Prosperity and Poverty*. New York: Crown Publishers, 2012.

Aitken, Jonathan. *Nazarbayev and the Making of Kazakhstan*. London: Continuum, 2009.

Arrow, K. "The Economic Implications of Learning by Doing." *The Review of Economic Studies* 29, no. 3 (1962): 155. http://dx.doi.org/10.2307/2295952

Asadov, M. et al. *Conserving Ecosystems of Inland Water Bodies in Central Asia and the South Caucasus. (Implementing the UN Millennium Development Goals in Central Asia and the South Caucasus: Goal 7: Ensure Environmental Sustainability)*. The Central Asian Regional Ecological Center, Almaty, Kazakhstan, 2006.

Asian Development Bank. *Country Partnership Strategy: Kazakhstan 2012–2016*, 2012.

Aslund, Anders. *Building Capitalism: The Transformation of the Former Soviet Bloc*. Cambridge, UK: Cambridge University Press, 2002.

Aslund, Anders. *How Capitalism Was Built: The Transformation of Central and Eastern Europe, Russia and Central Asia*. Cambridge, UK: Cambridge University Press, 2007.

Aslund, Anders. *Russia's Crony Capitalism: The Path from Market Economy to Kleptocracy*. New Haven, CT: Yale University Press, 2019.

Auty, R. "Resource-Driven Models of the Development of the Political Economy," in *Energy, Wealth and Governance in the Caucasus and Central Asia: Lessons Not Learned*, edited by R. Auty and I. de Soysa. New York: Routledge, 2006, pp. 17–26.

Auty, Richard M., and Indra de Soysa. eds. *Energy, Wealth and Governance in the Caucasus and Central Asia: Lessons not Learned*. London: Routledge, 2006.

Azerbaijan Fishery and Aquaculture Country Profiles. 2012. http://globserver.cn/eb/azerbaijan/fishery-and-aquaculture.

Beckert, Sven. "Emancipation and Empire: Reconstructing the Worldwide Web of Cotton Production in the Age of the American Civil War." *The American Historical Review* 109, no. 5 (December 2004): 1405–1438.

Beckert, Sven. *Empire of Cotton: A Global History*. New York: Vintage Books, 2014.

Beckert, Sven. "Empire of Cotton." *The Atlantic*. December 12, 2014.

Beyazli, D.S., Z.Y. Bayram, and S. Aydemir. "Policies on Population, Development and Pollution of the Black Sea Basin." *Scientific Research and Essays* 5, no. 12 (2010): 1418–1427.

Birnbaum, Michael. "Putin's Eurasian Economic Union Starts in 2015 with Curtailed Ambitions." *The Washington Post*. December 23, 2014.

Bissell, Tom. *Chasing the Sea: Lost Among the Ghosts of Empire in Central Asia*. New York: Pantheon Books, 2003.

BP. *BP Statistical Review of World Energy 2002–23*. London.

Breckle, Siegmar-W., Walter Wucherer, Liliya A. Dimeyeva, and Nathalia P. Ogar, eds. *Aralkum-A Man-Made Desert: The Desiccated Floor of the Aral Sea*. Heidelberg: Springer, 2012.

Brooks, Karen M. "Soviet Agricultural Policy and Pricing under Gorbachev," in *Soviet Agriculture: Comparative Perspectives*, edited by Kenneth R. Gray. Ames: Iowa State University Press, 1990, pp. 116–129.

Brown, Lester R. *US and Soviet Agriculture: The Shifting Balance of Power*. Washington, DC. Worldwatch Paper 51 (1982).

Chan, Samuel. "Pyrrhic Victory in the 'Tournament of Shadows': Central Asia's Quest for Water Security (1991–2009)." *Asian Security* 6, no. 2 (2010): 121–145.

Chen, Jiquan, et al. *Dryland East Asia: Land Dynamics Amid Social and Climate Change*. Beijing, China: Higher Education Press, 2013.

China's CNPC Imports First Gas from Kazakhstan ahead of Winter. *Reuters*. October 22, 2017.

Cohen, Ariel. *Kazakhstan: The Road to Independence – Energy Policy and the Birth of a Nation*. Washington, DC: Johns Hopkins SAIS/Silk Road Studies Program, 2008.

Cooley, Alexander, and John Heathershaw. *Dictators Without Borders: Power and Money in Central Asia*. New Haven, CT: Yale University Press, 2017.

Craumer, Peter R. "Trends in Soviet Dryland Farming and Soil Conservation Practices with Comparison to North American Developments," in *Soviet Agriculture: Comparative Perspectives*, edited by Kenneth R. Gray. Ames: Iowa State University Press, 1990, pp. 176–194.

European Union. Series: 1973–2002. *Environmental Action Programme* (6th EAP Covers 2002–2012). Brussels.

European Union. Commission. *Laying Down a Community Action Plan for the Conservation and Sustainable Exploitation of Fisheries Resources in the Mediterranean Sea under the Common Fisheries Policy*. Communication from the Commission to the Council and the European Parliament. COM 535 final, October 9, 2002. Brussels.

European Union. Commission. *Communication from the Commission on Rights-Based Management Tools in Fisheries*. February 26, 2007. Brussels.

European Union. Commission. *Review of the Management of Deep-Sea Fish Stocks*. Communication from the Commission to the Council and the European Parliament. COM 30 final, January 29, 2007. Brussels.

European Union. Council of Ministers. Council Regulation (EC) No 768/2005 of 26 April.

Establishing a Community Fisheries Control Agency and amending Regulation. (EEC) No 2847/93 Establishing a Control System Applicable to the Common Fisheries Policy, 2005.

Ferguson, Rob. *The Devil and the Disappearing Sea: A True Story about the Aral Sea Catastrophe*. Vancouver, BC: Raincoast Books, 2003.

Food and Agriculture Organization. *Fishery Country Profile: Turkmenistan*. FID/CP/TUK, August, 1998.

Food and Agriculture Organization. *Fishery Country Profile: Republic of Uzbekistan*. FID/CP/UZB, Rev. 1, January, 2003.

Food and Agriculture Organization. *Fishery Country Profile: Kazakhstan*. FID/CP/KAZ, March, 2004.

Food and Agriculture Organization. *Fishery Country Profile: Kyrgyz Republic*. FID/CP/KGZ, October, 2007.

Food and Agriculture Organization. *Inland Capture Fisheries and Aquaculture in the Republic of Uzbekistan: Current Status and Planning*. FAC/1030/1, Rome, 2009.

Food and Agriculture Organization. *Inland Fisheries Livelihoods in Central Asia: Policy Interventions and Opportunities*. Technical Paper 526, Rome, 2009.

Food and Agriculture Organization. *Review of the Current Status of Inland Capture Fisheries and Aquaculture in the Republic of Uzbekistan*. Part I. Rome. Conception of aquaculture and capture fisheries development of the Republic of Uzbekistan, 2008–2016. Part II. (Aquaculture and capture fisheries, development policy and strategy of the Republic of Uzbekistan, 2008–2016). Tashkent, 2009.

Food and Agriculture Organization. *Feasibility of Stocking and Culture-Based Fisheries in Central Asia*. FATP/565. Ankara, 2011.

Food and Agriculture Organization. European Commission on Agriculture. *Land Reform and Farm Performance in Europe: A 20-year Perspective*. Baku: Food and Agriculture Organization of the United Nations, 2012.

Food and Agriculture Organization. *Fisheries and Aquaculture in Tajikistan: Review and Policy Framework*. SEC/FIRA/C1030/3. Ankara, 2013.

Food and Agriculture Organization. *FAO Statistical Yearbook 2014: Europe and Central Asia-Food and Agriculture*. Regional Office for Europe and Central Asia. Budapest, 2014.

Fishelson, James. "From the Silk Road to Chevron: The Geopolitics of Oil Pipelines in Central Asia." *Geohistory*. School of Russian and Asian Studies. Winter, 2007.

Fracking Here, Fracking There. *The Economist*. 2011. http://www.economist.com/node/21540256

Goldman, Marshall I. *The Soviet Economy: Myth and Reality*. New York: Prentice-Hall, 1968.

Goldman, Marshall I. *USSR in Crisis: The Failure of an Economic System*. New York: W.W. Norton, 1983.

Goldman, Marshall I. *What Went Wrong with Perestroika*. New York: W.W. Norton, 1991.

Goldman, Marshall I. *Lost Opportunity: Why Economic Reforms in Russia Have not Worked*. New York: W.W. Norton, 1996.

Goldman, Marshall I. *The Piratization of Russia: Russian Reform Goes Awry*. New York: Routledge, 2003.

Goldman, Marshall I. *Petrostate: Putin, Power and the New Russia*. New York: Oxford University Press, 2008. (also published as Oilopoly by OneWorld Publications, 2008.)

Graham, Larissa J., and Brian R. Murphy. "The Decline of the Beluga Sturgeon: A Case Study About Fisheries Management." *Journal of National Resources and Life Sciences Education* 36, no. 1 (2007): 66–75.

Graham, Norman A. "The Prospect for Regional Governance of Inland Fisheries in Central Eurasia." Paper prepared for the FAO-MSU Global Conference on Inland Fisheries, January, Rome. Published in Select Proceedings of the Conference in 2016.
Graham, Norman A., and Cody Schulz. "The Political Economy of Economic Diversification in Azerbaijan, Kazakhstan and Russia: Progress and Challenges." The Great Steppe II: Papers presented to the Forum of Humanities of the International Turkic Academy. Astana, Kazakhstan. 9, 9 May, 2017, pp. 1–16.
Graham, Norman A., Stephen Pueppke, and Talgat Uderbayev. "The Current Status and Future of Central Asia's Fish and Fisheries: Confronting a Wicked Problem." *Water* (August, 2017).
Graham, Norman A., Stephen G. Pueppke, Sabir Nurtazin, Talgarbay Konysbayev, Farid Gibadulin, and Meiirli Sailauov. "The Changing Dynamics of Kazakhstan's Fisheries Sector: From the Early Soviet Era to the Twenty-First Century." *Water* 14, no. 9 (May, 2022): 1–22.
Gray, Kenneth R. *Soviet Agriculture: Comparative Perspectives*. Ames: Iowa State University Press, 1990.
Hornidge, Anna-Katharina, Anastasiya Shtaltovna, and Conrad Schetter, eds. *Agricultural Knowledge and Knowledge Systems in Post-Soviet Societies*. Bern: Peter Lang, 2016.
Johnson, Chalmers. *MITI and the Japanese Miracle*. Stanford, CA: Stanford University Press, 1982.
Johnson, D. Gale, and Karen McConnell Brooks. *Prospects for Soviet Agriculture in the 1980s*. Bloomington: Indiana University Press, 1983 (in Association with the Center for Strategic and International Studies, Georgetown University).
Jones Luong, Pauline, and Erika Weinthal. *Oil Is Not a Curse*. New York: Cambridge University Press, 2010.
International Crisis Group. *Central Asia: Water and Conflict*. ICG Asia Report No. 34. Osh/Brussels: May 30, 2002.
International Crisis Group. *The Curse of Cotton: Central Asia's Destructive Monoculture*. Asia Report No. 93. Bishkek/Brussels: 28 February 28, 2005.
Kaiser, Mark J., and Allan G. Pulsipher. "A Review of the Oil and Gas Sector in Kazakhstan." *Energy Policy* 35 (2007): 1300–1314.
Kandiyoti, Deniz, ed. *The Cotton Sector in Central Asia: Economic Policy and Development Challenges*. London: School of Oriental and African Studies, University of London, 2007.
Karimov, Islam. *Uzbekistan on the Threshold of the Twenty-First Century: Challenges to Stability and Progress*. New York: St. Martin's Press, 1998.
Kazakhstan. Message of President Nursultan Nazarbayev. *Prosperity, Security and Ever Growing Welfare of All the Kazakhstanis*. (7 Long Term Priorities and Agenda for 1998–2000), 1997.
Kazakhstan. Message of President Nursultan Nazarbayev. *Kazakhstan 2050 Strategy*. December 15 state of the nation address. Subsequent statements refined and expanded on the strategy with an emphasis on industrialization, entrepreneurship and human capital development, 2012.
Kazakhstan. Message of President Nursultan Nazarbayev. *Invest in Kazakhstan*, 2016.
Kazakhstan. Message of President Nursultan Nazarbayev. *Third Stage of Modernization in Kazakhstan: 5 Priorities*. January 31, 2017.
"Kazakhstan, Uzbekistan: Differing Approaches on Aral Sea." *Eurasianet*. March 22, 2012. http://www.eurasianet.ort/print/65167.

Kleveman, Lutz. *The New Great Game: Blood and Oil in Central Asia*. London: Atlantic Books, 2003.
"KMG Prepares for "Big Oil." *Oil and Gas of Kazakhstan* no. 5 (2012).
Kukeeva, Fatima, and Tolganay Ormysheva. "The Ili River: Is It a 'Casualty' of Kazakhstan's Complicated Relationship with China." Paper presented at the Annual Conference of the Central Eurasian Studies Society, University of Washington, Seattle. October, 2017.
Lavigne, Marie. *The Economics of Transition: From Socialist Economy to Market Economy*. Second edition. New York: St. Martin's Press, 1999.
Lederman, D., and W. Maloney. *Natural Resources, Neither Curse Nor Destiny*. Palo Alto, CA: Stanford Economics and Finance, an imprint of Stanford University Press, 2007.
Levine, Steve. *The Oil and the Glory: The Pursuit of Empire and Fortune on the Caspian Sea*. New York: Random House, 2007.
Lipovsky, Igor. "The Central Asian Cotton Epic." *Central Asian Survey* 14, no. 4 (1995): 529–542.
Litvin, Valentin. *The Soviet Agro-Industrial Complex: Structure and Performance*. Boulder, CO: Westview Press, 1987.
"LUKOIL Reduced the Oil Production in Kazakhstan by 4.7%." Oil and Gas of Kazakhstan no. 5 (2012).
Madramootoo, Chandra A., and Victor A. Dukhovny, eds. *Water and Food Security in Central Asia*. Dordrecht, Netherlands: Springer/NATO Science for Peace and Security Series-C: Environmental Security, 2011.
Marriott, James, and Mika Minio-Paluello. *The Oil Road: Journeys from the Caspian Sea to the City of London*. London: Verso, 2012.
Matley, Ian. "Central Asia XII. Economy in the 19th–20th Centuries." Encyclopaedia Iranica. V/2 (December 30, 2012).
Mazzucato, M. *The Entrepreneurial State*. New York: Anthem, 2013.
McCauley, Martin. *Khrushchev and the Development of Soviet Agriculture: The Virgin Land Programme 1953–1964*. New York: Holmes & Meier Publishers, Inc., 1976.
Medvedev, Zhores A. *Soviet Agriculture*. New York: W.W. Norton & Co., 1987.
Michel, Casey. "Tajikistan's Rogun Dam Rankles Uzbekistan." *The Diplomat*. July 21, 2016.
Micklin, Philip. "Desiccation of the Aral Sea: A Water Management Disaster in the Soviet Union." *Science* 241, no. 4870 (September 2, 1988): 1170–1176.
Micklin, Philip. *Managing Water in Central Asia*. London: The Royal Institute of International Affairs, Russia and Eurasia Programme, Central Asian and Caucasian Prospects, 2000.
Micklin, Philip, N.V. Aladin, and Igor Plotnikov, eds. *The Aral Sea: The Devastation and Partial Rehabilitation of a Great Lake*. Berlin: Springer, 2014.
Miller, Chris. *The Struggle to Save the Soviet Economy: Mikhail Gorbachev and the Collapse of the USSR*. Chapel Hill, NC: University of North Carolina Press, 2016.
Motamed, Mesbah J., et al. *Cotton and Hydropower in Central Asia: How Resource Competition Affects Trade*. Washington, DC: U.S. Department of Agriculture. Economic information Bulletin 106. January 2013.
Nazarbaev, Nursultan. *The Kazakhstan Way*. London: Stacey International, 2008.
Nikanova, Svetlana, and Tatyana Nazarova, *Aral Yesterday and Now: Problems and Perspectives of Aral's Crisis*. (photoalbum) Almaty, KZ: MFSA, 1997.
North, D. *Institutions, Institutional Change, and Economic Performance*. Cambridge, UK: Cambridge University Press, 1990.

OECD. *Review of Agricultural Policies: Kazakhstan*, 2013.

"Oil Production at Kashagan Planned to Start in Q1." *Oil and Gas of Kazakstan* no. 5 (2013).

Olcott, Martha Brill. *The Kazakhs*. Stanford, CA: Stanford University, Hoover Institution Press. Second edition, 1995.

Olcott, Martha Brill. *Central Asia's New States: Independence, Foreign Policy and Regional Security*. Washington, DC: U.S. Institute of Peace Press, 1996.

Olcott, Martha Brill. *Central Asia's Second Chance*. Washington, DC: Carnegie Endowment for International Peace, 2005.

Olcott, Martha Brill. *Kazakhstan: Unfulfilled Promise?* Washington, DC: Carnegie Endowment for International Peace, 2010.

Olcott, Martha Brill. *Tajikistan's Difficult Development Path*. Washington, DC: Carnegie Endowment for International Peace, 2012.

Overland, Indra, Heidi Kjaernet, and Andrea Kendall-Taylor, eds. *Caspian Energy Politics Azerbaijan, Kazakhstan and Turkmenistan*. London: Routledge, 2010.

Pannier, Bruce, and Edige Magauin. "Kazakhstan: China Discusses Future of Irtysh River." *RadioFreeEurope/RadioLiberty Archive*. May 9, 1999. https://www.rferl,.org/a/1091429.html.

Pannier, Bruce. "Turmenistan's Gas Gloom." *OilPrice.com*. March 11, 2017.

Pavlov, L. "Cotton in Russia." *The Russian Review* 1 (May 1916): 237–240.

Peimani, Hooman. *Failed Transition, Bleak Future? War and Instability in Central Asia and the Caucasus*. Westport, CT: Praeger, 2002.

Petr, Tomislav, and Valery P. Mitrofanov. "The Impact on Fish Stocks of River Regulation in Central Asia and Kazakhstan." *Lakes and Reservoirs: Research and Management* 3, no. 3 (1998): 143–164.

Peyrouse, Sebastien. "The Hydroelectric Sector in Central Asia and the Growing Role of China." *China and Eurasia Forum Quarterly* 5, no. 2 (2007): 131–148.

Pomfret, Richard. *Kazakhstan's 2030 Strategy: Goals, Instruments and Performance*. Paper presented at the American Economic Association annual conference, 4 January 2014.

Porter, Michael E. *Kazakhstan's Competitiveness: Roadmap Towards a Diversified Economy*. Harvard Business School, Institute for Strategy and Competitiveness, 2005.

Pueppke, Steven J., Sabir T. Nurtazin, Norman A. Graham, and Jiaguo Qi. "Central Asia's Ili River Ecosystem as a Wicked Problem: Unraveling Complex Interrelationships at the Interface of Water, Energy, and Food. *Water* 10 (April, 2018): 1–21.

Putz, Catherine. "Uzbekistan's Changing Rogun Tone." *The Diplomat*. July 10, 2017.

Putz, Catherine. "China's Imports of Turkmen Gas Decline." *The Diplomat*. December 27, 2017.

Robbins, Christopher. *Apples Are from Kazakhstan: The Land That Disappeared*. New York: Atlas and Co., 2008.

Rodrik, D., A. Subramanian, and F. Trebbi. "Institutions Rule: The Primacy of Institutions over Geography and Integration in Economic Development." *Journal of Economic Growth* 9, no. 2 (2004): 131–165. http://dx.doi.org/10.1023/b:joeg.0000031425.72248.85

Ross, M. *The Oil Curse: How Petroleum Wealth Shapes the Development of Nations*. Princeton, NJ: Princeton University Press, 2012.

Roy, Olivier. *The New Central Asia: The Creation of Nations*. New York: New York University Press, 2000.

Sachs, J., and A. Warner. "Natural Resource Abundance and Economic Growth." *NBER Working Paper Series*. Working Paper 5398 (1995). http://dx.doi.org/10.3386/w5398

Sarymbetova, Aygerim. "Kazakhstan to Start Gas Deliveries to China." *Caspian News*. October 10, 2017.

Shapiro, Sidney. (ed.), *Our Changing Fisheries*. Washington, DC: US Department of Commerce, 1971.

Siegelbaum, Lewis. "Great Fergana Canal." *Seventeen Moments in Soviet History*. National Endowment for the Humanities-MSU Matrix. 1939. http://soviethistory.msu.edu/1939-2/great-fergana-canal/.

Spechler, Martin C., Joachim Ahrens, and Herman W. Hoen. *State Capitalism in Eurasia*. Singapore: World Scientific Publishing Co., 2017.

Starr, S. Frederick, and Svante E. Cornell. (eds.), *Uzbekistan's New Face*. Lanham, MD: Rowman & Littlefield, 2018.

Stucki, Virpi, Kai Wegerich, Muhammad Mizanur Rahaman, and Olli Varis. (eds.), *Water and Security in Central Asia: Solving a Rubik's Cube*. London: Routledge, 2014.

The Aral Sea Crisis. http://www.columbia.edu/tmt2120/ and http://www.envis.maharashtra.gov

Thorpe, Andy, et al. "The Collapse of the Fisheries Sector in Kyrgyzstan: An Analysis of Its Roots and Its Prospects for Revival." *Communist and Post-Communist Studies* 42 (2009): 141–163.

Treadgold, Donald W. *The Great Siberian Migration: Government and Peasant in Resettlement from Emancipation to the First World War*. Princeton, NJ: Princeton Legacy Library, 1957.

U.S. Department of Agriculture. Foreign Agricultural Service. Global Agricultural Information Network. *Cotton and Products Annual: Republic of Uzbekistan*. Report UZ2023-0001. April 13, 2023.

U.S. Department of Agriculture. Foreign Agricultural Service. International Production Assessment Division. *Cotton Explorer*.

Utegenova, Ailuna R. *Kazakhstan's 2030 Development Strategy: Significance and Results*. 2020. https://ifsh.de/file-CORE/documents/yearbook/english/10/Utegenova-en.pdf

Uzan, Marc. (ed.), *The New Kazakhstan: Steering the Steppe Amid a Global Storm*. Paris: Reinventing Bretton Woods Committee, 2023.

Uzbekistan, Ministry of Agriculture. *Progress Report on the Last Quarter Century*. May, 2018.

Weiss, Linda. *The Myth of the Powerless State*. Ithaca, NY: Cornell University Press, 1998.

Westerman, Frank. *Engineers of the Soul: The Grandiose Propaganda of Stalin's Russia*. New York: The Overlook Press, 2002.

Wheeler, William. *Environment and Post-Soviet Transformation in Kazakhstan's Aral Sea Region*. London: UCL Press, 2021.

Whitman, John. "Turkestan Cotton in Imperial Russia." *The American Slavic and East European Review* 15, no. 2 (April, 1956): 190–205.

Wittfogel, Karl A. *Oriental Despotism: A Comparative Study of Total Power*. New Haven, CT: Yale University Press/New York: Vintage Books, 1957/1981.

World Bank. *Kazakhstan: Transition of the State*. Washington, DC. 1997.

World Bank. Europe and Central Asia Region. *Water Energy Nexus in Central Asia: Improving Regional Cooperation in the Syr Darya Basin*. Washington, DC. 2004.

World Bank. *Assessment Studies for Proposed Rogun Hydropower Project in Tajikistan*. Washington, DC. September 1, 2014.

World Bank. *World Development Indicators*. Annual. http://data.worldbank.org/data-catalog/world-development-indicators

World Bank and Food and Agriculture Organization. *Market Growth Potential for Kazakhstan Fishery and Aquaculture Products.* July 2022.

Zhanbosinova, Albina, Fatima Kukeyeva, Tolganay Ormysheva, and Norman A. Graham. "Irtysh Whirls: The Geopolitics of River Rights Between Kazakhstan, China and Russia." *CERES Occasional Papers*. Center for European, Russian and Eurasian Studies. Michigan State University, July, 2019.

Zinzani, Andrea. *The Logics of Water Policies in Central Asia: The IWRM Implementation in Uzbekistan and Kazakhstan.* Zurich, Switzerland: LIT Verlag, 2014.

Zonn, Igor S., and Audrey G. Kostianoy (eds.), *The Turkmen Lake Altyn Asyr and Water Resources in Turkmenistan.* Berlin: Springer-Verlag, 2014.

5

RUSSIA AND CENTRAL EURASIA

How Strong Is Russia's Hold on Its "Near Abroad"?

The end of the USSR essentially came with the resignation of Mikhail Gorbachev on December 25, 1991. This followed numerous consultations and maneuvering on Gorbachev's part in order to hold some semblance of the Soviet "Empire" together. In the end, there was no prospect for keeping the three Baltic republics—Estonia, Latvia, and Lithuania—in the union, and as Serhii Plokhy argues in rich detail, the independence drive of Ukraine's leadership, perhaps most notably Leonid Kravchuk, proved irresistible. Independence movements in Moldova and the Caucasus further tipped the balance. As Plokhy stressed:

> ...while the lost arms race, economic decline, democratic resurgence, and bankruptcy of communist ideals all contributed to the Soviet implosion, they did not predetermine the disintegration of the Soviet Union. That was caused by the imperial foundations, multiethnic composition, and pseudofederal structure of the Soviet state, features whose importance was fully recognized neither by American policy makers in Washington nor by Gorbachev's advisers in Moscow.[1]

One must also add the crucial role of Boris Yeltsin after his successful resistance to the attempted coup in August of 1991. His calculated efforts to marginalize Gorbachev's desperate actions to preserve a Union, and his willingness to work with Kravchuk in support of an independent Ukraine were decisive. Kravchuk and Yeltsin then joined Stanislav Shushkevich of the Republic of Belarus to sign an agreement to establish the Commonwealth of Independent States (CIS) on December 8 at a hunting lodge in Belavezha, Belarus, effectively ending Gorbachev's campaign for a new Union and dissolving the USSR. The CIS never really functioned as anything like the USSR or the replacement Union that Gorbachev envisioned. It was supplemented though by the Collective Security Treaty Organization (CSTO)

DOI: 10.4324/9781003373735-5

composed of Armenia, Belarus, Kazakhstan, Kyrgyzstan, the Russian Federation, and Tajikistan as members. Peacekeeping forces were engaged on several occasions, notably in Kyrgyzstan and Tajikistan. Most recently, Russian forces were deployed in January 2022, when President Tokayev of Kazakhstan requested assistance to curtail violent protests stimulated by a substantial increase in fuel costs.

Armenia was clearly disappointed by the lack of CSTO response to its recent conflicts with former member Azerbaijan in 2022 and 2023. Armenia had invoked Article 4 of the treaty, comparable to NATO's Article 5, requiring a response by the members if one is attacked. Russia was undoubtedly preoccupied with Ukraine, but the refusal to support the Armenian side in effect meant that the Turkish support for Azerbaijani operations seemed to be decisive. Armenian politicians actually raised the prospect that membership in NATO might be more effective. There was some speculation that Putin was in fact punishing the Armenian leadership for its in-depth bilateral consultations with President Macron of France.

We should also note that the Shanghai Cooperation Organization (SCO) was formed by the People's Republic of China, the Russian Federation, Kazakhstan, Kyrgyzstan, Tajikistan, and Uzbekistan; it then grew to include India and Pakistan as members. Afghanistan, Belarus, Iran, and Mongolia have observer status with some expectation to become full members. Initially established in 1996 as the Shanghai 5 to promote mutual confidence building for border security in the region of newly independent, post-Soviet states, it added Uzbekistan in a proclamation in St Petersburg on June 15, 2001, and the agreement entered into force in September 2003 with a focus on anti-terrorism cooperation. Some inter-bank and infrastructure development cooperation has evolved since, and additional members are under consideration.

The concept of Eurasianism fit nicely with Putin's eventual effort to strengthen his influence over former Soviet republics and territories after he succeeded Yeltsin in 2000. Kazakhstan's founding President Nursultan Nazarbayev had suggested the formation of a framework for closer economic cooperation among post-Soviet states. Putin embraced the suggestion and formed the Eurasian Economic Union (EaEU) which he boasted would ultimately replicate the level of economic integration achieved by the European Union. The EaEU became Putin's instrument to try to bring Ukraine back under Russian influence and stem the potential attraction of the European Union to many Ukrainians. The EU's neighborhood initiative and its subsequent invitation for Ukraine to sign an Association Agreement were threatening to Putin's ambitions. Indeed, under Putin's pressure, President Viktor Yanukovych subsequently opted to join the EaEU instead of the EU Association Agreement. This led to widespread opposition among pro-EU Ukrainians in Kyiv's Maidan Square. President Yanukovych and his security forces mishandled the response with provocative violence, and he ultimately was forced to flee to Russia.

As noted earlier, Putin chose to use this as an opportunity to annex Crimea and foment separatism in the Eastern Donbas region of Ukraine.[2] In hindsight, the

February 2022 Russian invasion of Ukraine was a logical extension (at least in his own mind) of Putin's plan to recover empire.

This plan was fairly evident by 2012. The prospect for collaboration with the West to address challenges of global terrorism was dim by then, and there appeared little Russian interest in efforts to deal more broadly with global poverty and instability, let alone climate change, in a constructive manner. Januz Bugajski and Margarita Assenova provided a somewhat depressing assessment of all of this in *Eurasian Disunion* in 2016:

> The war between Russia and Ukraine has dramatically escalated the geostrategic competition between the Western states and a neo-imperial Russia. Moscow's [2014] invasion of Ukraine and annexation of Crimea demonstrates President Putin's geopolitical ambitions and the limitations of Western deterrents. It has challenged the independence of other nearby post-Soviet states and exposed America's newest NATO allies, NATO partners, and even non-aligned European countries to the destabilizing regional repercussions of the Kremlin's assertiveness. Although Russia's military capabilities do not match those of its Soviet predecessor, the country presents a destabilizing presence in several key regions and employs numerous forms of subversion against targeted neighbors.[3]

Indeed, Bugajski and Assenova detailed 68 distinct tools and methods of an "Arsenal of Subversion" from diplomatic pressures, border challenging, innovative propaganda and espionage, and exploitation of frozen conflicts that have been in play in the near abroad, as they reviewed five regional flanks of Soviet successor states and former clients.

How far Putin's plan for recovering the empire will be pursued and to what extent he has been able to maintain or extend his influence over the "near abroad" is the subject of this chapter. What are the roots of this, and to what extent does it derive from personal drive and domestic considerations? And to what extent does this frustrate or otherwise influence environmental sustainability in the larger Eurasian region?

The Roots of Russian Eurasianism and Imperialism

The expansion of the Russian Empire to include the Caucasus, Northeastern Europe, the North shore of the Black Sea, and much of Central Asia is a good place to start, as one recalls the imperial exploits of Tsars and especially Peter the Great, culminating in 1721 after the Great Northern War, and Catherine the Great after 1762. Alexander I's success in defense of the invasion by Napoleon helped to consolidate Russia as a European great power. Rivalries with the British, Hapsburg, Ottoman, Persian, Chinese, and Japanese empires were the preoccupation of Russian foreign policy after the 240 years of Mongol rule. Imperial conflict continued in large measure until the 1917 Revolution and World Wars of the Twentieth Century. After the allies defeated Nazi Germany, the Soviet Union consolidated its own empire, with the lands of most of the furthest reaches of the Russian Empire.[4] As Serhii Plokhy has stressed, this multinational/multiethnic empire was ultimately untenable, as the Center weakened politically and economically.[5] Of course, the Soviets would deny the charge of imperialism because only capitalists can be imperialists! But by 1991, the Marxist argument found little support as an assessment of the reality facing the Soviet Union.[6]

Plokhy argued:

> What seems so special about the Soviet Union is that very few people considered it an empire during its lifetime or were prepared to treat it as anything but a nation-state....
>
> Like the British dominions, the Soviet republics left their metropolis in 1991 under the leadership of their own "native" leaders and institutions. As was true of other twentieth-century dominions and colonial possessions, some of the Soviet republics left the Union core not against the wishes of the dominant nation but in accordance with them: the leaders of the Russian Federation wanted the Central Asians to go once Ukraine left the Union. Also, as in the case of other European empires, it was the question of extending citizenship rights, particularly voting rights, to residents of the Soviet republics that made the continuation of the empire in its existing form all but impossible.[7]

President Yeltsin had his hands full trying to reform and rule the political economy of the new Russian Federation. As a consequence, he was not very effective in maintaining influence over the other successor states to the USSR. Western, Chinese, and Turkish influence grew while he was preoccupied with a divided and obstreperous Duma. As noted in Chapter 4, independent pipelines were constructed to take oil and natural gas from Azerbaijan, Kazakhstan, and Turkmenistan to Europe, Iran, and China, effectively ensuring some degree of economic independence.

Yeltsin's economic and political challenges were complicated by national self-determination pressures in the Federation. Chechnyan leaders, in particular, became adamant in their drive for independence, and serious hostilities began in December 1994. The war was devastating to both sides, but Yeltsin found it necessary to seek a ceasefire in 1996 and signed a peace treaty finally in 1997. This was not a heroic outcome to a war deeply unpopular for Russians. The peace treaty was not enough, and when Vladimir Putin was picked to serve as Yeltsin's Prime Minister, he employed a different approach to eliminating the resistance from Chechnya which erupted again in 1999; Chechnya was not finally subdued until 2009 after horrible devastation. Putin was relentless, and he rallied considerable domestic support after the Chechens were accused of deadly apartment bombings in three Russian cities.

There were Russian critics of Putin's handling of Chechnya, of course. For example, in *A Dirty War,* Anna Politkovskaya offered penetrating questions of the official strategy and its justification:

> To begin with, I thought how senseless everything happening here was. If you look at it from the State's point of view, why scatter a vast number of mines around the city and receive in return an astronomic growth in the number of disabled people, who require tons of medicine, artificial limbs and so on? And then scatter more mines. And again ferry in medicine, etc. Now it's clear what the State is up to. Its concern for the situation is purely virtual; the only reality is the scattering of mines. No matter how much we want to believe the reverse, or attribute everything to our chronic disorder or thieving, the reality is that the inhabitants of Grozny have been sentenced to this fate. Evidently, the ultimate aim is to ensure that as many people in the city as possible are either left without legs – or dead.[8]

Masha Gessen was also critical of Putin's prosecution of the Second Chechnyan War. She visited Grozny and reported on the devastation and civilian suffering. Gessen also provided detail on Politkovskaya's treatment during and after her journalistic service in Grozny and its environs—subjected to a mock execution, poisoned on a flight to Beslan to observe the vicious and bungled school attack purportedly by Muslim extremists, and ultimately murdered on her Moscow doorstep.[9]

Putin was indeed unrelenting in his efforts to pacify the Chechens. Suspicions about whether the motivating bombings of Russian apartment buildings in Buynaksk, Moscow, and Volgodonsk in September 1999 were in fact orchestrated by FSB agents were reinforced subsequently by the discovery of an explosive device in the city of Ryazan.[10] This was later passed off as a security training exercise, and Putin ordered air strikes on Grozny on September 24.

Alexander Litvinenko and Yuri Felshtinsky amassed considerable evidence against the view that the attacks were by Chechens and argued in *Blowing Up Russia* that this was a "Plot to Bring Back KGB Terror"- and foment war in Chechnya, presumably as a way to raise Putin's popularity in the eyes of nationalistic Russians.[11] Litvinenko was later poisoned with Polonium and died in London on November 23, 2006. The Fourth Edition of his book with Felshtinsky not only trumpeted that the book was banned in Russia, but declared it was "The Bestseller that got Litvinenko Assassinated"!

Western reaction to Putin's campaigns in the Second Chechen War was somewhat divided, at least initially. In part, this reflected the legacy of concern about the global threat of terrorism, especially that perceived as growing from Wahhabist Islamic fundamentalism, but there was also sentiment in support of the Russian government's effort to maintain the territorial integrity of the Russian Federation. Part of this legacy undoubtedly harkened back to sentiments that led President Clinton to compliment Boris Yeltsin's own campaign against the Chechnyan drive for national self-determination as akin to Abraham Lincoln's posture during the Civil War: "that no state had a right to withdraw from our union. And so, the United States has taken the position that Chechnya is a part of Russia." Presidents Clinton and Yeltsin faced serious electoral tests in 1996 and saw advantage in playing down differences.[12]

President Putin's support of President Bush after the Osama Bin Laden inspired attacks on the United States on September 11, 2001, also helped to downplay sentiment against his Chechnyan campaign.[13] Part of this dialogue included discussion of NATO enlargement and the prospect of strategic cooperation between the United States and Russia. The NATO-Russia Council was established at the NATO-Russia Summit in Rome on May 28, 2002, with a declaration on "NATO-Russia Relations: A New Quality." The groundwork for this had been provided in the 1997 NATO-Russia Founding Act on Mutual Relations.[14]

Putin reportedly was the first head of state to telephone President Bush after the 9–11 attacks, subsequently supporting actions to bring down the Taliban regime in Afghanistan, and search for Bin Laden. What followed was rare and valuable overflight and logistics support granted by Russia, enabling more efficient supply of the US and allied forces on the ground in the region. The Russians did not stand in the way of US military bases established in Kyrgyzstan and Uzbekistan as part of the logistics effort.

Putin's strong words at the 2007 Munich Security Conference reflected a serious change in his attitude toward the role of the United States in nearby regions, especially Iraq:

> We are seeing a greater and greater disdain for the basic principles of international law. And independent legal norms are, as a matter of fact, coming increasingly closer to one state's legal system. One state and, of course, first and foremost the United States, has overstepped its national borders in every way. This is visible in the economic, political, cultural and educational policies it imposes on other nations. Well, who likes this? Who is happy about this?[15]

The Obama administration did try to "reset" US-Russian relations, based on clear recognition that Russia's cooperation was needed for meeting major objectives, and that the Bush administration had allowed the relationship to stagnate, mostly through neglect. As US Ambassador to Moscow, Michael McFaul, described it, the Obama administration posture for the reset was to determine and pursue shared interests, develop a multidimensional relationship, avoid any temptation to try to isolate or coerce Russia, increase engagement with Russian society, and display respect for Putin and his accomplishments in revitalizing Russia. However, US values would not be downplayed, and US relations with Russia's neighbors, most notably Georgia and Ukraine would not be sacrificed. A "benign" regional environment was necessary if there was to be a chance for Russian democratization.[16]

The timing of the reset had appeared optimal given that Dmitry Medvedev was President. Putin had stepped aside in line with the constitutional prohibition against more than two consecutive terms. He was certainly not out of the picture, however, as Prime Minister (and President-in-waiting, as it turned out), but Medvedev seemed more accommodating to US initiatives and appeared to have a more globalist perspective on the best path to a prosperous future for Russia.[17]

Obama and Medvedev worked together on a number of important initiatives, including overflight support for the war in Afghanistan and concerns about the threat from a potential nuclear-armed Iran.[18] Looking back on it, Putin was remarkably tolerant of Medvedev's posture on close relations with the United States and Europe (at least publicly), but his time to return to the Presidency was approaching; and Putin saw personal motivations for establishing his control and his plan for reclaiming lost empire that did not envision subordination to the United States. Accordingly, McFaul regarded 2010 as the high point of the reset progress. It was all downhill from there.

Catherine Belton argued that Putin was intentional in his efforts to challenge the West from the outset, returning Russia to great power status.[19] Joshua Yaffa captured the impact eloquently, noting that Crimea's "return" in 2014 "...hit the right notes for the Russian national psyche, a measure of revenge for all the slights and insults of the post-Soviet era."[20]

What then guided Russian foreign policy toward Post-Soviet Europe and Eurasia? Was it Russian nationalism together with imperial ambition? Writing in 2016, Charles Clover traced the rise of Russia's new nationalism in the context of Eurasianism, in *Black Wind, White Snow.*[21] In Clover's account, a central figure in this was Alexander Dugin who first built a case for a new nationalist orientation in *The Foundations of Geopolitics* in 1997.[22] Dugin's subsequent books added a strong inclination to turn back the tide of Western liberalism. The argument flowed from his vision of a long-term geopolitical struggle between the civilization of the Sea (especially the United States), and the civilization of the Land (especially Russia), and led to an irresistible goal that Russia must at minimum dominate the north shore of the Black Sea. With Ukraine in the way of this ambition, one could foresee the geopolitical conflict between Russia and the West that grew dramatically in 2014.

Dugin became an advisor to President Putin and subsequently developed his geopolitical ideology. The clearest statement of his view of Russian Geopolitics was in *Last War of the World-Island: The Geopolitics of Contemporary Russia.*[23] Here, he delineated the fundamental features of the civilization of the Land: (1) conservatism; (2) holism; (3) collective anthropology (the *narod* is more important than the individual); (4) sacrifice; (5) an idealistic orientation; and (6) the values of faithfulness, asceticism, honor, and loyalty. "Beyond ideological preferences, Russia is doomed to conflict with the civilization of the Sea."[24]

In *Eurasian Mission: An Introduction to Neo-Eurasianism*, Dugin pointed to globalization (originating in the West), as a challenge to the nations and civilizations of the Eurasian continent. His conception of the Eurasian movement was "…strongly against globalization as a form of ideological, economic, political, and value-based imperialism.…The nations of Eurasia must be free and independent."[25] In important respects, Dugin saw liberalism with its focus on individualism as the key threat.

Dugin subsequently argued for a Fourth Political Theory, superseding the first (liberalism), the second (Communism/socialism); and the third (fascism/National Socialism). In his words: "The 4PT considers itself non-modern or counter-modern." It emphasized the importance of history and tradition, as well as the centrality of the problem of identity.[26] From the point of view of the 4PT, totalitarianism is connected with liberal ideology—the third totalitarianism:

> Liberalism is a totalitarian ideology that insists, through the classical methods of totalitarian propaganda, that the individual has the highest value.…
>
> Liberal society, which puts itself in opposition to the collective societies of socialism and fascism, has itself become a collective – a standardized and stereotypical one. The more an individual aspires to be unique within the context of the liberal paradigm, the more he becomes similar to everyone else. Liberalism brings with itself the stereotyping and homogenization of the world, which destroys all forms of diversity and differentiation.[27]

The United States was simply a "conceptual society" that lacked a real tradition. Dugin identified the United States then as a "Country of Absolute Evil."[28] The 21st century will be defined by the conflict between Eurasianists and Atlanticists.[29] Indeed, his view was that the Atlanticists want

> …to impose a single model of economic order on all peoples in the world, elevating the experience of the economic development of the Western part of world civilization in the nineteenth and twentieth centuries to the status of a universal standard.[30]

Dugin's animosity toward the United States and the West in general led him to embrace extreme right nationalist groups in Europe and the United States as potential allies.[31]

Dugin did not seem to have a positive economic strategy, beyond curtailing globalization. He offered nothing that would promote broad-based prosperity in Russia and the Eurasian successors to the Soviet Union. He made no effort to defend the old Soviet system's centrally planned economy, or indeed any sort of statist recipe. Indeed, he was preoccupied with defending Eurasia from the West and preserving Eurasian traditions and "differentialism."

He was not sanguine about the emerging state of the Russian economy, however. In *Putin vs Putin*, he stressed the need to develop a "real economy" in Russia, but he provided no details on what must be done to match or surpass the diversified manufacturing and high technology economies of the East Asia and the West.[32] His only clear intervention on the challenges and prospect of developing high technology industries was to denigrate Medvedev who had supported the Skolkovo Innovation Centre project (which had been developed with MIT participation and became the Skolkovo Institute of Science and Technology).[33]

While Dugin served as an advisor to Putin, especially in the early years of the Putin Presidency (when there appeared to be a search for a justifying ideology), he became disenchanted with Putin's indecisiveness and his apparent willingness to accept the liberal globalizing paradigm. Commentary on this view was offered in an interview format at the end of his *Putin vs. Putin* book.[34] Dugin developed impatience with Putin, as have some of his more aggressive critics on the conduct of the invasion of Ukraine, but elements of his arguments in Putin's speeches and interviews can still be identified.

Writing in 2008, Marlene Laruelle offered an interesting and detailed exposition of Neo-Eurasianism arguments in *Russian Eurasianism: An Ideology of Empire*; she concluded:

> Thus (Neo-) Eurasianism lives in the virtual world of a Eurasian identity that has always been more complex than the Eurasianists conceived and of a political unity that is no more. However, it is not only a political doctrine; it sees itself as the extension of the Russian imperial imagination, and it awkwardly formulates the "mental atlas" of citizens who consider themselves to have lost out in the changes that have taken place in this part of the world over the past twenty years. So is Eurasia a hyperbolic version of an alternative identity; the dream of a *different* Russia?[35]

Dugin owed a debt for his Eurasianist perspective to the writings of Lev Gumilev, often regarded as the true father of Neo-Eurasianism. Mark Bassin offered an excellent delineation of these ideas and their impact in *The Gumilev Mystique*. Gumilev is honored in Russia and Central Asia, especially Kazakhstan. His influence in the late 1980s and 1990s placed emphasis on socio-biological explanations of nationality and ethnicity (ethnos and ethnogenesis) and on the deterministic historical development of Russia. His early writings on the Gokturk Khaganates and on the Rus and the Golden Horde established his scholarly reputation and have been drawn upon to help construct notions of Russian geopolitical connections to Central Asia. He has been criticized for Anti-Semitism and for writings that others use for exclusionary arguments against foreign ethnic groups; he was bitterly anti-Western in his assessment of the collapse of the USSR and his rejection of the West's "so-called universal human values,"[36] but as Bassin concluded:

> ....Gumilev spoke and speaks today to a diverse variety of constituencies – from the Russian president to fundamentalist insurgents – and offers a diverse (and contradictory) array of explanations for the most important social and national

> issues....It is not too much to say that Lev Gumilev has become a universal point of reference and metadiscourse in his own right: a venerable and apparently inexhaustible wellspring of ideas and inspiration for pretty much anyone seeking to make sense of the Russian past, present, and future.[37]

The ideology of Eurasianism is less attractive, beyond Russia and Central Asia—at least less attractive than was Marxism—and may be less prized than the simple value of the importance of preserving order. This has some resonance with the precept, often attributed to Islamic and/or Chinese roots: "Better Sixty Years of Tyranny than One Night of Anarchy."[38] Freedom is less important than order.

In the Russian context, it can also be connected with the arguments of Ivan Alexandrovich Ilyin, the early 20th century Russian religious philosopher who, in the context of the Russian Civil War, called for "Resistance to Evil by Force" and warned against chaotic doctrines and movements. He stressed concern about (democratic?) doctrines that "...were legitimizing weakness, exalting egocentrism, indulging a lack of will, robbing the soul of its social and civic duties ... prone to simplifying a naively idyllic world outlook." The common argument that Russians largely desire a strong leader of a strong state, especially since the chaotic experiment with democracy under Yeltsin, is relevant. Indeed, Putin has made this claim in speeches, quoting Ilyin's anti-democratic arguments (especially in defense of the Russian Orthodox Faith).[39]

Can these notions drive the long-term project of committed authoritarian leaders? Is legitimacy important to a leader like Putin? Does it matter from where or how it is derived? Does truth matter? James Heiser's religious/ideological interpretation of Dugin's Eurasianism suggested that it should but does not. War was inevitable. "The American Empire Should Be Destroyed."[40]

Both Eurasianism and the value of order put some blame on the impact of Western inspired globalization for the ills besetting post-Communist Russia. What is interesting, however, is the extent to which this argument ignored what happened in Yeltsin's faulty transition Russia, and indeed the more complicated dynamics of what had happened globally, good and bad.[41]

Eurasianism offers a convenient justification for aggressive actions in Georgia, the Baltics, and especially Ukraine. But his actions have already been counterproductive in alarming Western countries and reaffirmed, indeed expanded NATO commitments and national rearmament.

James J. Sheehan's argument in *Where Have All the Soldiers Gone* that Europe was transforming into a "Civilian Super-State" had a certain reality through much of the post-World War II rise of the European Community and Union. Some European governments were especially eager to take advantage of the expected "peace dividend" after the end of the Cold War and essentially disarm after the collapse of the Soviet Union. This added to the objective reality that NATO was, at best, a paper defensive alliance and posed no real offensive threat to Russia. NATO enlarged, but offensive force projection capacity was minimal and there were clear reductions in NATO member defense expenditure and readiness.[42] This began to change with Putin's military build-up and naval and air operations in the Baltic Sea and Black Sea regions, and of course the interventions in Georgia and Ukraine. Putin's fixation on the threat from the West while ignoring (at least in public pronouncements) the growth of China as a rival to the East, especially in the "near abroad" regions of Central Eurasia has seemed puzzling and counterproductive.

One could argue with some confidence that Russia's security in fact declined as a result of Putin's post-2007 initiatives, with Russia now facing challenge from the East, West, and South.[43] Putin in effect has stimulated a reluctant European rearmament and encouraged a better prepared alliance to its West while ignoring a potentially rising power to the East and Southeast. Was this an unintended consequence or the result of a calculation that the domestic payoffs from a more assertive and interventionist Russia outweighed the potential security drawbacks? Do these geopolitical calculations derive from Eurasianism or develop in spite of it?

Catherine Belton painted a dark and aggressive picture of Russian Foreign Policy under Putin, citing the influence of the former KGB cronies that protect and promote Putin's posture of challenging the liberal world order.[44] Russia's engagement in Syria since September 2015, at the request of President Assad, has been expensive in lives and treasure, but the payoff in gaining a foothold in the Middle East, helping to counter ISIS terrorism, and confronting the United States and perhaps Turkey geopolitically may be worth the expense to Putin. As the United States retreated, it almost seemed like Russia in Syria was re-enacting the old struggle for the region between the Russian and Ottoman empires.[29,45]

Dmitri Trenin posited *The End of Eurasia* in 2002 given the impact of globalization and the geopolitical situation in which a weakened Russia found itself. But he saw some alternative scenarios for the region. For Russia, he saw three alternatives: (1) Revisionist Russia; (2) Russia's Disintegration; or (3) Creative Adjustment. Revisionist Russia or the "reimperialization" of Russia, as Henry Kissinger described it, was favored by traditional elites, but was unlikely to succeed according to Trenin.[46] For Russia's neighbors, he saw significant challenges and uncertainty, and given that "Revisionist Russia" was a serious option to consider, no matter how difficult, the need was clear for the independent states created out of the collapse of the USSR to proceed intentionally and carefully. The success or failure of the various strategies employed remains very much in question.

Resistance to Neo-Imperialism in the Near Abroad

Chapter 4 provides considerable detail on the sustainable development ambitions of several of the successor states to the former Soviet Union. The Baltic States of Estonia, Latvia, and Lithuania are the obvious success stories, as they have pursued effective political and economic transition policies to enable rapid accession to the European Union and to join EU efforts to respond to climate change and fossil fuel dependence.[47] Obviously, these efforts were given added stimulus by the Russian invasion of Ukraine.

The Republic of Georgia was modestly successful in its efforts to embrace Western values of liberal democracy and market capitalism, especially after the "Rose Revolution" led by Mikheil Saakashvili in 2003, but the reforms did not take hold as firmly as in the Baltics.[48] The Russian intervention in 2008, illustrating the insecurity of the neighborhood, brought Saakashvili's efforts to a close in many respects. But the Western orientation has remained strong. Saakashvili's early economic reform successes, and his work to root out blatant corruption ultimately seemed to fade, as one began to detect his unilateral impulses and the white-collar corruption that continued to infect the economy.

Putin in effect trapped Saakashvili into an overconfident push to regain control of the two separatist territories of Abkhazia and South Ossetia. This led to an embarrassing military defeat at the hands of the Russians. The Georgians were forced to face Russia without Western help—evidence of Saakashvili's wishful thinking about a NATO response. NATO leaders, especially cautious European ones, were unwilling to disrupt relations with Russia, particularly when some believed Saakashvili's actions were reckless.[49]

After a series of scandals, the Georgian Dream Party, orchestrated by Bidzina Ivanishvili, a former Russian oligarch and billionaire, was victorious. This spelled the end of Saakashvili as a serious competitor in Georgian politics. Georgian Dream subsequently prevailed in elections ever since in a bitterly divided political context. Opposition politicians and mass protestors frequently criticize what they perceive as a lean toward Russia, but the actual policies of the government have reflected a delicate effort to balance pro-Western interests with a desire not to provoke an aggressive Russian response. The pattern of Russian trade policy toward Georgia exemplifies the reward and punishment of efforts to control the South Caucasus. Trade with Russia was curtailed, for example when the Georgia Dream government signed an association agreement with the EU.[50]

The economic ties between Georgia and Russia have persisted and even grown under the Western sanctions regime, and Russian refugee flows to Georgia have been substantial. But Georgians overwhelmingly support eventual membership in the European Union and may yet achieve candidate status along with Moldova and Ukraine. Despite its ties to Russia, Georgia has voted in favor of UN resolutions condemning Russia over the invasion of Ukraine and is tacitly in support of Western sanctions.[51]

Some Russian media outlets outlawed since the Russian invasion of Ukraine has transferred their operations to Tbilisi. The Georgian Dream Government sparked dramatic protests when it introduced a "foreign influence" bill that was reminiscent of similar legislation adopted in Russia with clear repressive goals. The target in Russia was critical domestic media and "foreign agents." The Georgian protests against the more benign Georgian bill had a dramatic impact, and it was withdrawn.[52] This decision seems to support the view that the stance of the Government has been and remains somewhat ambiguous. On the one hand, it does not support the Russian invasion, but on the other, it is not ready to work aggressively against it.

On May 16, 2023, Georgia announced the resumption of direct flights to Russia in contravention of the EU ban on flights to Russia. Whereas the step to withdraw potential regulation of foreign influence was seen as a step toward EU candidacy status, the resumption of direct flights to Russia was unlikely to be warmly received in Brussels. Some critics argue that the Georgian political system needs a new model of democracy, given the current divisions and single-party dominance.[53]

Belarus and Ukraine were less successful than Georgia in their push for autonomy, for different reasons, but the latter seems intent on an independent development strategy if it can rid itself of Putin's imperialistic aggression. Writing in 1999, Yuri Fedorov and Ghia Nodia reflected on the Russian elite perspective on the claims to continuing control of Central Eurasia:

> There are, of course, various roots of the integrationist attitudes, more properly called imperial attitudes, that are characteristic of a substantial part of the Russian elite; they range from the native perceptions of some leading political figures who see themselves as masters of a great power—which Russia is no longer—to cynical manipulations of nostalgic feelings as part of a domestic political power struggle. But whatever its origins, the implementation of an integrationist strategy in practical actions and policies is already a source of insecurity and instability in the Southern Tier. For Russia itself, reintegration would inevitably turn into an enormous burden that is unbearable for its ruined economy, and lead to a confrontation with the United States and other powers, as well as stimulating suspicions and hostility among most of the local elites.[54]

There are indeed burdens replete in assuming full imperial responsibility for Central Eurasia. Reducing unsustainable use of scarce water resources, for example, in diversifying agriculture away from the cotton monoculture, especially in Uzbekistan, has been a challenge, that independent governments in the region have begun to confront. Similarly, reducing the over-dependence on fossil fuel exports in Azerbaijan and Kazakhstan has been addressed to some extent, and Turkmenistan has also joined these two in diversifying its customer base and transit options.

Alexander Cooley provided an interesting slant on this challenge with respect to the external influence of Russia, China, and the United States, especially since 9–11, in *Great Games, Local Rules: The New Great Power Contest in Central Asia*. In his words,

> All of the Central Asian States have learned to play the great powers off one another for their local benefit. However, their exact tactics and demands have depended on their institutional structures, their capacity, and the natural resource endowments bequeathed to them as independent states.[55]

This seemed to be somewhat successful when Russia was distracted by its own "transition challenges" under Yeltsin and in the early years of Russian-US-Chinese interest in regional stability and anti-terrorism efforts after 9–11. One must ask if independent Central Asian strategies are really viable now with a more insistent and politically motivated Russia and a cash-rich China eager to extend its influence and economic power.

Writing in 2006, Lena Jonson sketched a systematic explanation as to why Russia under Putin had changed from the traditional Soviet and Russian policy to maintain exclusive influence over Central Asia. International weakness was prominent in her explanation, but there was also some willingness on the part of Putin to seek collaboration with the United States and China on important global issues, and domestic economic factors, including energy supply ties, were relevant. She posited that the world was changing rapidly, and Russia had had to adapt to a less central role in world politics, including less exclusivity in Central Asia.[56] Russia's vicious and burdensome operations in Ukraine likely reduce the tools and capability to return to exclusive influence in the Eurasian region, but the inclination is undoubtedly still important in the calculus of Putin and his henchmen.

The challenge of resisting undue Russian influence is now more complicated than it was when Cooley made his case on the impact of "local rules," but there may be more opportunities for developing stronger external relationships and domestic strengths, thanks to the Ukraine quagmire into which Russia has fallen. Central Asian governments can diversify their economies and economic partners given the foundations they have laid and the apparent weakness of a Russia that is unable or at least unwilling to be so actively engaged in the region. On the other hand, several of these governments seemed to have found ways to take advantage of infrastructure development funding through the Belt and Road Initiative (BRI) and investments through the Chinese centered Asia Infrastructure Investment Bank (AIIB) without falling into the "debt trap" that has plagued several African countries, as well as South and Southeast Asian countries, most notably Sri Lanka. Central Asia enjoys more active trade, especially imports, with China than with Russia these days. The energy exports to China discussed in Chapter 4 enable balanced trade. Chinese and Western products are plentiful in the markets. Chapter 6 will explore the Chinese relationship with Central Eurasia in detail.

A word of caution here is prudent. Keir Giles, writing in *Moscow Rules: What Drives Russia to Confront the West*, published in 2019, was much more pessimistic about the recent and future role of Russia in Eurasia and beyond, pointing to long historical roots and characteristics distinctive from that of Western powers:

> Understanding and accepting this basic clash of interests and worldviews inevitably means that the West must invest heavily and for the long term in deterring Russia from aggressive and hostile actions. This deterrence must include willingness to impose costs and consequences on Russia in response to military adventurism, cyber and information assaults, or any of the other ways in which Russia endangers the integrity of Western states or the lives and livelihoods of their citizens.[57]

What does this mean for the successor states to the Former Soviet Union and the Warsaw Treaty Organization. What is required to ensure independence for these governments in the face of the history of Russian expansion and revanchism?

In the view of Giles:

> President Putin himself is not the issue. Enduring Russian concerns, rather than Putin's personal preferences, are the drivers for Russia's current policy priorities, and his successor (if and when one might appear) is highly unlikely to abandon altogether the strain of conservation nationalism that serves Russia well today. Putin is enacting, rather than inventing, Russian state assumptions and ambitions that have been consistent over the long term and remained unchanged even during the period when Russia was treated as a friend and partner of the West. What has changed dramatically in the last decade is Russia's capability to achieve these ambitions.[58]

It seems clear that there is little popular interest among the successor states to the Soviet Union to fully re-integrate. That does not mean that there is motivation for severing ties completely with the Russian Federation. But the record of Russian reluctance, when the leadership had more influence and power, to help thaw "frozen conflicts" in Nagorno-Karabakh between Armenia and Azerbaijan, between Georgia and Russia over South Ossetia and Abkhazia, over Transnistria between Moldova and Russia has been modest to say the least. Moreover, the border disputes between Uzbekistan-Kyrgyzstan and Kyrgyzstan-Tajikistan and a range of trans-border water sharing disputes appear well off Russia's radar screen at present.

President Putin reportedly expressed his determination that Russia would never leave the South Caucasus. In a speech before the Russian-Armenian Interregional Forum held in Gyumri, Armenia, in December of 2013:

> On the contrary, we will make our place here even stronger. We will strengthen our position here, drawing on the best of what our forebears left us and with the support of good relations with all countries in the region, including Armenia.[59]

One can see evidence of this in his effort to sabotage Azerbaijan's energy pipelines to Europe through Turkey, with South Stream and to keep the Nagorno-Karabakh conflict frozen.[60] But the South Stream strategy collapsed in part due to reaction to the annexation of Crimea.

One can also argue that the Nagorno-Karabakh conflict is closer to "resolution" than it has been since the two parties, Armenia and Azerbaijan became independent in 1991. There was no question that ethnic Armenians resided in the disputed territory, but there is almost no international recognition that Armenia should have legal claim to the region, or that the Republic of Artsakh or Nagorno-Karabakh Republic has legal claim to govern it. The United Nations General Assembly adopted resolution No. 62/243 on March 14, 2008 calling for complete withdrawal of Armenian forces from Azerbaijani territory.

The Russian Federation figured prominently in ceasefire efforts, having been given some responsibility for resolving the dispute in the context of the OSCE Minsk Group, co-chaired by France, Russia, and the United States, but the conflict remained frozen with little movement until the 2020 war changed the facts on the ground. Azerbaijani military forces recovered most of its territory, with substantial technical assistance from Turkey, and the Minsk Group in effect became irrelevant. Tension between Russia and the West over Ukraine had further limited what modest role the Group could play. Delegations from Armenia and Azerbaijan met regularly in Brussels with EU mediators since 2020.[61]

There is some argument that Russia and Central Asia are drifting apart. As Temur Umarov of the Carnegie Endowment for International Peace argued in December of 2022, "...Russian soft power in Central Asia is dissipating before our eyes." He went on to argue, however, that the Russian invasion of Ukraine has led to a growth in trade in the region, as Central Asian and Caucasus states have taken advantage of Russian agricultural import and energy export needs, in the wake of sanctions.[62] In addition, trade from EU and other countries sanctioning trade to Russia grew to Central Asian and Caucasus states which in turn saw an increase in exports (re-exports) to Russia.[63]

As noted earlier, Russia has been highly motivated for most of its post-Soviet experience to maintain as much control as possible of the successor states in Central Asia. It has sought to do this by economic ties and dependence as well as political-military relationships. The Central Asian region is characterized, much like Crimea and Eastern Ukraine, by substantial ethnic Russian residents and Russian military ties and installations. Accordingly, the Governments in question must tread carefully as they seek to display serious concern about the invasion of Ukraine and their own territorial integrity and independence, as well as to solve their own regional environmental and sustainable development challenges.

Leadership change in Uzbekistan has made a difference in specific regional relationships, in the wake of less Russian engagement. For example, there is the path breaking agreement signed between Uzbek President Shavkat Mirziyoev and Kyrgyz President Sadyr Japarov during a state visit in Bishkek, January 26–27, 2023. This appears to be an indication of emerging independent diplomacy to solve long standing border and water management disputes, while Russia was MIA. The agreement included moving the Kempir-Abad reservoir to Uzbek management and committed Uzbek help to Bishkek to build the hydroelectric plant called Kambarata-1.[64]

More generally, S. Frederick Starr and Svante E. Cornell collected a series of important chapters in *Uzbekistan's New Face*, to document a series of domestic and foreign policy initiatives instituted by the Mirziyoev administration. Richard Weitz, for example, argued that the change from Karimov meant a softening in Uzbekistan's opposition to Tajikistan's dam projects, including the controversial (massive) Rogun Dam now finally close to completion. In some respects, the Dam was too far along to be stopped, but there was some Uzbek hope that water sharing arrangements could be supported.[65]

Initiated in 1976 as a Soviet project, the Rogun Dam on the Vakhsh River has been tormented by financial and political problems since the outset. It was essentially abandoned in 1993 after the collapse of the USSR. Serious construction was resumed by the Tajik Government in 2016, despite opposition from President Karimov, who feared that it would threaten the water supply needed for what was widely regarded as unsustainable irrigation of the vast Uzbek cotton fields.[66] Weitz documented a series of reforms meant to make the Uzbek economy more competitive globally, as well as more sustainable. There seemed to be some emphasis on moving on from the suspicion that President Karimov harbored for Russian intentions. For example, Uzbek energy purchases from Russia expanded in 2018–2020. But there was also considerable attention focused on improving economic and political relations with the United States, the EU, and China.[67]

Of course, one should recall that Karimov laid a foundation of independence from Moscow early in his tenure as the foundation president of the Republic of Uzbekistan after 1991. He was suspicious of both Russian imperial aims and Western democratization and transition market economic policies. Accordingly, he was cautious in adhering to Russian sponsored security and economic integration plans and institutions. In and out of the CSTO and unwilling to support the Russian position in the UN on the annexation of Crimea, Uzbekistan has suffered Moscow's verbal wrath at times, but both post-Soviet Uzbek presidents have been skillful in resisting Russian leverage by firmly defending state sovereignty and building economic relationships with China and the West.[68]

There was hope for some time in Tajikistan that Russia would help with the Rogun Dam financing, especially during the period of what Michael Klare called the Russian "Energy Juggernaut" of very high energy export income, up to the 2008 financial crisis, but nothing really came from this.[69] The World Bank financed a feasibility study of the completion plans in 2018. World Bank Vice President Cyril Muller indicated at the launch ceremony that:

> Rogun can be a source of much needed, inexpensive, renewable and reliable energy for the people of Tajikistan. It can also be an important source of energy exports that can generate revenues for other critical investment in human capital to help transform Tajikistan's economy and create the foundation for a prosperous future.[70]

The World Bank then largely left it to the Tajik Government to self-finance the dam gradually. An Italian construction firm, Salini Impregilo, was contracted for the construction at $3.9 billion. In January of 2023, the World Bank provided funding for technical assistance to help implement the project.

While Russian relations have remained stronger with President Emomali Rahmon of Tajikistan than perhaps with any other Central Asian leader, given his longevity in power since 1994, there are clear limits to what Russia can now do economically and militarily with the invasion of Ukraine under way. But remittances from Tajiks working in Russia have long been an important component of the economic relationship between the two countries.

Turkmenistan is the Central Asian successor state with clearly the most secure independence from post-Soviet Russian influence. The decisions to establish independent markets and pipeline infrastructure for Turkmen natural gas in Iran and especially China are part of the explanation, but the clear isolationist postures from Russian-inspired regional security and economic organizations pursued by Turkmenbashi and his successors were determinative. Turkmenistan did not join the CSTO, the SCO, or the EaEU. Unfortunately, most of the Turkmen population did not prosper from the resulting isolation, given the harsh authoritarianism, corruption, and inequality. There remains little sign of reform aimed at improving broad-based prosperity or responsible governance. As discussed in Chapter 4, this lack of responsible governance extends to environmental lapses. Unsustainable use of water continues unabated, and there is little evidence that the Turkmen government is taking serious action to eliminate the methane gas leaks from two gas fields that continue to contribute mightily to greenhouse gases. One report suggested that the leaks contributed more to global warming in 2022 than did all of the United Kingdom's carbon emissions.[71]

Kazakhstan, under President Kassym-Jomart Tokayev, however, is a different story. The reform efforts presented have been substantial and promising in the wake of his designation to replace Nursultan Nazarbayev, who formally resigned as President in 2019.[72] There was concern that Tokayev would abandon his predecessor's "multi-vector" foreign policy and align even more closely with Russia after the Russian/CSTO intervention to support him in the violent inflation-related unrest of January 2022. The subsequent Russian invasion of Ukraine one month later, however, seemed to renew an inclination to be more independent, albeit with some caution. Kazakhstan has not voted in support of the invasion in the context of the various United Nations resolutions, and it has been a somewhat willing recipient of Ukrainian and Russian refugees fleeing the violence and Russian military conscription.

Speaking with Putin publicly at the St. Petersburg International Economic Forum in June 2022, "Tokayev made it very clear that his country remained committed to the principle of territorial integrity and therefore would not recognize the 'quasi-states' of Donetsk and Luhansk."[73] A similar approach had been taken by Kazakhstan in 2014—not recognizing the Russian annexation of Crimea, despite the long, close relationship that Putin and Nazarbayev had enjoyed. Kazakhstan formally abstained in UNGA resolution 68/262 (27 March 2014) condemning the annexation, and was not among the few states that subsequently recognized the annexation as legal. Only Armenia and Belarus among all of the post-Soviet successor states recognized the annexation. Tokayev has consulted with President Zelensky on several occasions and has provided humanitarian relief to Ukraine. On February 16, 2023, Tokayev reportedly conveyed his support for "a diplomatic solution to the Russia-Ukraine conflict based on the United Nations Charter and universally accepted international law principles."[74] Clearly, he has remained committed to the principle of territorial integrity, which of course fits nicely with Kazakhstan's own historic concerns about Russian influence, particularly in the north where the ethnic Russian population has been substantial.

Nonetheless, Kazakhstan has remained a member of all of the Russian-inspired and dominated regional organizations, especially the CSTO and the Eurasian Economic Union, but it has diversified its economic relationships well beyond them, particularly with China, Turkey, Saudi Arabia, and Iran, but also the EU and the United States. Some of this is quite rational to implement policies in place for perhaps the last ten years to diversify the economy both away from its dependence on oil and gas exports generally and toward more environmentally sustainable agricultural production, given water constraints.[75]

But one can also detect some concern about the long-term Russian agenda for the region, beyond Ukraine, as well as an unsurprising response to heavy handed Russian efforts to influence the political economy of Kazakh energy fortunes. Russia suspended the use of its pipeline network for Kazakh oil shipments four times during 2022, likely in response to the Kazakh government's support of Ukraine's territorial integrity.[76] On the other hand, Kazakh exports to Russia have grown since the war began and Western sanctions grew. Some of this appears to have been due to the rise of transshipment strategies by traditional Russian trading partners seeking to reach Russia through Kazakhstan. The Kazakh Government denies this, but "Sanctions Evasion" does seem to be a feature of the trade patterns of Central Eurasian countries. Indeed, Russia has been developing a transportation corridor with Iran to boost bilateral trade and to ease shipments to and from India as well.[77]

Kazakhstan has been a popular destination for refugees from Russia—both those initially disturbed or disadvantaged by the original invasion decision and those subsequently threatened with conscription by the partial mobilization announced September 21, 2022 to replace the large number of troops lost in the fighting with Ukraine. But fear of the large numbers of migrants, including some with possible pro-Russian nationalist sentiments among them led to a December 31, 2022 Kazakh policy amendment limiting the visa-free travel between Kazakhstan and other Eurasian Economic Union countries to no more than 90 days within a year without an employment contract or official justification. Such justification could include family reunification and educational enrollment.[78]

Kazakhstan has also been plagued off and on by currency instability relating to financial and trade ties with Russia. Western sanctions in response to the annexation of Crimea, not to mention the much more comprehensive and burdensome ones levied after the Russian invasion of Ukraine in 2022, have dramatically weakened the Ruble. This in turn made Russian exports to Kazakhstan very cheap. The more general failure of the EaEU to live up to Putin's promises has encouraged the Kazakhs to seek broader economic ties and alternative markets.[79]

Conclusion

Despite the pervasive Russian propaganda on television channels widely available in most Central Asian and Caucasus media markets, public sentiment in support of Ukraine has grown steadily, admittedly with some generational differences, given divergence in the type of media consumed (internet and social media-based vs. television).[80] Political elites, however, in most cases have remained cautious in public statements, UN votes, and concrete actions beyond reducing Russian language education and changing street designations.

The Central Eurasian states which have been most independent in their positions on the Russian invasion of Ukraine, especially Kazakhstan, but also Azerbaijan and the Republic of Georgia, and of course isolationist Turkmenistan, probably cannot be expected to be more aggressive in their dissent from Russian policies at this stage. The pro-Ukrainian alliance should be concerned about the latent Russian influence at work and the fragility of the political economies in question, making them potentially vulnerable to destabilization.

It is difficult to anticipate how the Russian military operations in Ukraine will end, but the stalemate and horrific losses of soldiers and equipment are undoubtedly unsettling to the Kremlin. One can envision more prospect for independence and perhaps environmentally sustainable policies and programs among the USSR successor states in Central Asia and the Caucasus if the Russian Federation continues to decline economically and remain dysfunctional politically.

On the other hand, as Tatiana Stanovaya, stressed in a May 2023 interview by Isaac Chotiner printed in *The New Yorker*, the opposition dissent from Putin's prosecution of war includes a radically "patriotic" segment, more aggressive than Putin, that favors total mobilization and a full war footing. They are "…incrementally dragging the country into a more repressive state."[81] Such an eventuality may lead to even more aggression in the near abroad. This suggests the need for vigilance and deterrent strategies along the lines of those suggested by Keir Giles. Retired Brigadier General Kevin Ryan's recent warning on "Why Putin Will Use Nuclear Weapons in Ukraine" makes this need for vigilance and strategy shockingly imperative.[82]

One should also consider the points made by Thomas Goltz in his revised and expanded version of *Georgia Diary.* Goltz drew a comparison between the early strategies for autonomy of Azerbaijan under Ilham Aliyev and Georgia under Mikheil Saakashvili:

> While Tbilisi embraced an openly anti-Russian policy in virtually all spheres –"baiting the bear" is one way to describe it –Baku was going out of its way to reassure Moscow of the long history that had bound the two fraternal peoples together, as part of a great tactical schmooze-job, and one that apparently has worked (at least so far).[83]

Georgia suffered military intervention and reduced prospect for retaining sovereignty over Abkhazia and South Ossetia and adopted a more cautious strategy under the Georgian Dream governments. Azerbaijan seems to have been able to develop a more independent position in its cautious engagement with NATO's Partnership for Peace and economically (remaining outside of the EaEU) with the help of substantial Western investment to produce, market, and transport its oil and gas to Europe.

What will work for Central Eurasia going forward, especially in light of the massive impact of the Russian invasion of Ukraine? If a void develops in Russia's ability to control its "Near Abroad," will China or Turkey help to fill it?[84] Or will "local rules" prevail? Chapters 6 and 7 explore this in detail.

Notes

1 Serhii Plokny, *The Last Empire: The Final Days of the Soviet Union* (New York: Basic Books, 2014), p. xvii.
2 Paul D'Anieri, *Ukraine and Russia: From Civilized Divorce to Uncivil War*, Second Edition (Cambridge: Cambridge University Press, 2023).
3 Januz Bugajski and Margarita Assenova, *Eurasian Disunion: Russia's Vulnerable Flanks* (Washington, DC: The Jamestown Foundation, June 2016), p. 57.
4 Central Asia, for the purposes of this volume, encompasses the USSR successor states of Kazakhstan, Kyrgyzstan, Tajikistan, Turkmenistan, and Uzbekistan. References to Central Eurasia, by custom, adds the USSR successor states of Armenia, Azerbaijan and Georgia to the other five.
5 Plokhy, *The Last Empire*, pp. 400–406. See also: James P. Duffy and Vincent L. Ricci, *Czars: Russia's Rulers for Over One Thousand Years* (New York: Barnes & Noble, 1995); John Keep, *Last of the Empires: A History of the Soviet Union 1945–1991* (Oxford: Oxford University Press, 1995); and Dominic Lieven, *Empire: The Russian Empire and Its Rivals* (New Haven, CT: Yale University Press, 2001).
6 This is a longstanding debate in the literature on imperialism, particularly among those who venture into the writings of Marx and Lenin. A sample of this writing that fits somewhat with Putin's critique of the unipolar excesses of US foreign policy after the end of the Cold War is offered by anthropologist David Harvey in *The New Imperialism* (Oxford: Oxford University Press, 2003). See also: Vladimir Lenin, *Imperialism, The Highest Stage of Capitalism* (1916).
7 Plokhy, *The Last Empire*, pp. 393–394.
8 Anna Politkovskaya, *A Dirty War: A Russian Reporter in Chechnya* (London: The Harvill Press, 2001), pp. 218–219.
9 Gessen, *The Man Without a Face*, pp. 145–154 and 212–218.
10 Ibid., pp. 216–217. See also: Alexander Litvinenko and Yuri Felshtinsky, *Blowing Up Russia: The Secret Plot to Bring Back KGB Terror* (New York: Encounter Books, 2007).
11 Litvinenko and Felshtinsky, *Blowing Up Russia*, pp. 3–30.
12 Terrence Hunt, "Clinton, Yeltsin Trade Compliments, Play Down Differences." *Associated Press News*, April 21, 1996.
13 http://www.ljubljana-summit.gov.si/en/index.html.
14 https://nato.int/cps/en/natohq/topics_50091.htm
15 Vladimir Putin, *Speech to the Munich Conference on Security Policy*, February 10, 2007. http://en.kremlin.ru/events/president/transcripts/24034.
16 McFaul, *From Cold War to Hot Peace*, chapters 6 and 7.
17 Ibid., p. 116.
18 Ibid., pp. 102–106.

19 Catherine Belton, *Putin's People: How the KGB Took Back Russia and Then Took on the West* (New York: Farrar, Straus and Giroux, 2020).
20 Joshua Yaffa, *Between Two Fires: Truth, Ambition, and Compromise in Putin's Russia* (New York: Tim Duggan Books, 2020), p. 209.
21 Charles Clover, *Black Wind, White Snow: The Rise of Russia's New Nationalism* (New Haven, CT: Yale University Press, 2016).
22 Alexander Dugin, *The Foundations of Geopolitics*. Russian Edition (Moscow: Arktogia, 1997).
23 Alexander Dugin, *Last War of the World-Island: The Geopolitics of Contemporary Russia* (London: Arktos, 2015).
24 Ibid., pp. 7–11.
25 Ibid., p. 39.
26 Ibid., pp. 115–116.
27 Ibid., p. 107.
28 Ibid., p. 145.
29 Ibid., see pp. 110–126.
30 Ibid., p. 64.
31 Ibid., p. 168.
32 Alexander Dugin, *Putin vs. Putin: Vladimir Putin Viewed from the Right* (London: Arktos, 2014), p. 226.
33 Ibid., p. 192.
34 Ibid., pp. 178–179.
35 Marlene Laruelle, *Russian Eurasianism: An Ideology of Empire* (Baltimore, MD: The Johns Hopkins University Press, 2008), p. 221.
36 Mark Bassin, *The Gumilev Mystique: Biopolitics, Eurasianism, and the Construction of Community in Modern Russia* (Ithaca, NY: Cornell University Press, 2016), pp. 211–214.
37 Ibid., p. 316.
38 Noah Feldman, "Better Sixty Years of Tyranny than One Night of Anarchy," Luncheon Speech delivered to Loyola Law School and published in *Loyola of Los Angeles International and Comparative Law Review* 143 (2009): p. 143.
39 Ivan Alexandrovich Ilyin, *On Resistance to Evil by Force* (London: Taxiarch Press, 2018) (originally published privately in 1925, Berlin), p. 6. See also: *Our Tasks*. Volume 1 (a collection of his articles 1948–1951 published together in Russian in 2014, Pubmix.com); and Timothy Snyder, "Putin's Philosopher of Russian Fascism." *The New York Review*, April 5, 2018. https://www.nybooks.com/daily/2018/03/16/ivan-ilyin-putins-philosopher-of-russian-fascism/

Adiscussion of the impact of Ilyin's writings on Putin in Catherine Belton, *Putin's People*, pp. 259 and 272.
40 James D. Heiser, *"The American Empire Should Be Destroyed": Alexander Dugin and the Perils of Immanentized Eschatology* (Malone, TX: Repristination Press, 2014), pp. 95–126.
41 See: Anders Aslund, *Russia's Crony Capitalism: The Path from Market Economy to Kleptocracy* (New Haven, CT: Yale University Press, 2019); Steven Radelet, *The Great Surge: The Ascent of the Developing World* (New York: Simon & Schuster, 2015); and Joseph E. Stiglitz, *Globalization and Its Discontents – Revisited: Anti-Globalization in the Era of Trump* (New York: W.W. Norton & Co., 2018).
42 James J. Sheehan, *Where Have All the Soldiers Gone: The Transformation of Modern Europe* (New York: Houghton Mifflin Harcourt, 2008). See also: U.S. Congress. Office of Technology Assessment, *After the Cold War: Living with Lower Defense Spending*. OTA-ITE-524, February 1992.
43 On the prospect for Russian rivalry and partnership with China, see: Paul J. Bolt and Sharyl N. Cross, *China, Russia and Twenty-First Century Global Geopolitics* (UK:

Oxford University Press, 2018); and Michal Lubina, *Russia and China: A Political Marriage of Convenience—Stable and Successful* (Opladen, Germany: Barbara Budrich Publishers, 2017).

44 Belton, *Putin's People*, especially Parts Two and Three.

45 See: Henry Foy, Laura Pitel an Chloe Cornish, "Erdoğan and Putin Announce Ceasefire in Idlib." *Financial Times*, March 5, 2020.

46 Dmitri Trenin, *The End of Eurasia: Russia on the Border Between Geopolitics and Globalization* (Washington, DC: Carnegie Endowment for International Peace, 2002), pp. 305–318.

47 Romuald Misiunas and Rein Taagepera, *The Baltic States: Years of Dependence 1940–1990* (London: Hurst and Co., 1993).

48 See: Irakly Areshidze, *Democracy and Autocracy in Eurasia: Georgia in Transition* (East Lansing, MI: Michigan State University Press, 2007).

49 See: Ronald D. Asmus, *A Little War That Shook the World* (New York: Palgrave Macmillan, 2010); and Thomas Goltz. *Georgia Diary: A Chronicle of War and Political Chaos in the Post-Soviet Caucasus* (Armonk, NY: M.E. Sharpe, 2009).

50 Bugajski and Assenova, *Eurasian Disunion*, p. 290.

51 Alexander Atasuntsev, "Is Georgia's Ruling Party Really Pro-Russian? *Carnegie Endowment for International Peace*, May 2, 2023. https://carnegieendowment.org/politika/89655.

52 Tony Barber, "Divided Georgia searches for elusive political stability." *Financial Times*, March 18, 2023.

53 Stephen Jones and Natalie Sabanadze, "Elections Are Not Enough: Georgia Needs a New Model of Democracy." *Eurasianet.org Perspectives*, March 10, 2023.

54 Yuri E. Fedorov and Ghia Nodia, "Conclusion: Regional Security for Russia, the Caucasus, and Central Asia—Going from Zero-Sum to Positive-Sum Policies," in *Russia, the Caucasus, and Central Asia: The 21st Century Security Environment*, eds. Rajan Menon, Yuri E. Fedorov, and Ghia Nodia (Armonk, NY: M.E. Sharpe, 1999), p. 251.

55 Alexander Cooley, *Great Games, Local Rules: The New Great Power in Central Asia* (Oxford University, 2012), p. 10.

56 Lena Jonson, *Vladimir Putin and Central Asia: The Shaping of Russian Foreign Policy* (London: I.B. Taurus, 2006), pp. 198–199.

57 Keir Giles, *Moscow Rules: What Drives Russia to Confront the West* (London: Chatham House, 2019), p. 173.

58 Ibid., p. 161.

59 Bugajski and Assenova, *Eurasian Disunion*, p. 292.

60 Ibid., pp. 290–291.

61 Heydar Isayev, "Armenian and Azerbaijani Leaders Meet for the Fifth Time in Brussels." *Eurasianet.org*, May 16, 2023.

62 Temur Umarov, "Russia and Central Asia: Never Closer, or Drifting Apart? Carnegie Endowment for International Peace," December 12, 2022. https://carnegieendowment.org/politika/88698.

63 Maxim Chupilkin, Beata Javorcik and Alexander Plekhanov, "The Eurasian Roundabout: Trade Flows into Russia through the Caucasus and Central Asia," *EBRD Working Paper No. 276,* European Bank for Reconstruction and Development, February 2023.

64 Chris Rickleton, "For Kyrgyzstan and Uzbekistan 'No Other Path' But Deeper Cooperation." RFE/RL, January 31, 2023.

65 Richard Weitz, "Uzbekistan's New Foreign Policy: Change and Continuity Under New Leadership," in *Uzbekistan's New Face*, eds. S. Frederick Starr and Svante E. Cornell (Lanham, MD: Rowman and Littlefield, 2018), p. 57.

66 See: Deniz Kandiyoti (ed.), *The Cotton Sector in Central Asia: Economic Policy and Development Challenges* (London: School of Oriental and African Studies, 2007), for detailed discussion of the complexity of this sector and its importance to Uzbekistan, as well as to Kazakhstan and Turkmenistan.

67 Weitz, *Uzbekistan's New Foreign Policy*, pp. 60–67.
68 Bugajski and Assenova, *Eurasian Disunion*, pp. 409–420.
69 Michael T. Klare, *Rising Power, Shrinking Plant: The New Geopolitics of Energy* (New York: Metropolitan Books, 2008).
70 World Bank, Press Release No: 2019/ECA/61; and Press Release No: 2023/55/ECA.
71 "Turkmenistan: Fiddling While Gas Leaks." Eurasianet.org, May 16, 2023. https://eurasianet.org/turkmenistan-fiddling-while-gas-leaks. See also: https://www.theguardian.com/world/2023/may/09/mind-boggling-methane-emissions-from-turkmenistan-revealed.
72 For example, see: Kazakhstan, *Adoption of Concept of Civil Society Development until 2025*. https://adilet.zan.kz/rus/docs/U2000000390; and *Law on Peaceful Assemblies*. https://strategy 2050.kz/ru/news/zakon-o-mitingakh-chto-izmenilos-za-god-/
73 Marie Dumoulin, "Steppe Change: How Russia's War on Ukraine Is Reshaping Kazakhstan," *European Council on Foreign Relations Policy Brief*. Ecfr.eu, April 2023, p. 3.
74 Ibid.
75 Norman A. Graham, et al., "The Changing Dynamics of Kazakhstan's Fisheries Sector: From the Early Soviet Era to the Twenty-First Century," *Water* (May 2022).
76 Dumoulin, "Steppe Change," p. 5.
77 Vigal Chazan, "How Viable Is Russia and Iran's 'Sanctions-Evasion' Corridor?" *Geopolitical Monitor*, April 6, 2023.
78 Dumoulin, "Steppe Change," p. 6.
79 Bugajski and Assenova, *Eurasian Disunion*, pp. 405–408.
80 Ibid., pp. 6–9.
81 Isaac Chotiner, "Why Russian Elites Think Putin's War Is Doomed to Fail." *The New Yorker*, May 3, 2023, p. 7.
82 Kevin Ryan, "Why Putin Will Use Nuclear Weapons in Ukraine." *Russia Matters*, May 17, 2023. https://w.w.w.russiamatters.org/analysis/why-putin-will-use-nuclear-weapons-in-ukraine.
83 Thomas Goltz, *Georgia Diary: A Chronicle of War and Political Chaos in the Post-Soviet Caucasus* (Armonk, NY: M.E. Sharpe, 2009), p. 258.
84 See: James Palmer, "Is China Replacing Russia in Central Asia? *Foreign Policy*, May 23, 2023. https://foreignpolicy.com/2023/05/23/china-central-asia-xi-xian-summit-russia-ukraine/?tpcc=recirc_latest06291; and Meaghan Tobin, "China Sends a Subtle Message to Central Asia: Rely on Us, Not Russia." *Washington Post*, May 19, 2023. https://www.washingtonpost.com/world/2023/05/19/china-central-asia-russia-summit/

References

Abdygaliyeva, Gauhar, et al. *Economic Diversification in the Republic of Kazakhstan Through Small and Medium Enterprise Development: Introducing New Models of Finding for SMEs*. New York: Columbia University, School of International and Public Affairs, 2007.

Aitken, Jonathan. *Nazarbaev and the Making of Kazakhstan*. London: Continuum, 2009.

Anker, Morten, et al. *The Caspian Sea Region towards 2025: CASPIA Inc. National Giants or Trade and Transit?* Delft, Netherlands: Eburon, 2010.

Areshidze, Irakly. *Democracy and Autocracy in Eurasia: Georgia in Transition*. East Lansing: Michigan State University Press, 2007.

Asadov, M., et al. *Conserving Ecosystems of Inland Water Bodies in Central Asia and the South Caucasus (Implementing the UN Millennium Development Goals in Central Asia and the South Caucasus: Goal 7: Ensure Environmental Sustainability)*. Almaty, Kazakhstan: The Central Asian Regional Ecological Center, 2006.

Aslund, Anders. *Russia's Crony Capitalism: The Path from Market Economy to Kleptocracy*. New Haven, CT: Yale University Press, 2019.

Auty, R. "Resource-Driven Models of the Development of the Political Economy," in *Energy, Wealth and Governance in the Caucasus and Central Asia: Lessons Not Learned*, edited by R. Auty and I. de Soysa. New York: Routledge, 2006, pp. 17–26.

Auty, Richard M., and Indra de Soysa (eds.), *Energy, Wealth and Governance in the Caucasus and Central Asia: Lessons Not Learned.* London: Routledge, 2006.

Beckert, Sven. "Emancipation and Empire: Reconstructing the Worldwide Web of Cotton Production in the Age of the American Civil War." *The American Historical Review* 109, no. 5 (2004): 1405–1438.

Belton, Catherine. *Putin's People: How the KGB Took Back Russia and Then Took on the West*. New York: Farrar, Straus and Giroux, 2020.

Beyazli, D.S., Z.Y. Bayram, and S. Aydemir. "Policies on Population, Development and Pollution of the Black Sea Basin." *Scientific Research and Essays* 5, no. 12 (2010): 1418–1427.

Birnbaum, Michael. "Putin's Eurasian Economic Union Starts in 2015 with Curtailed Ambitions." *The Washington Post*. December 23, 2014.

Chan, Samuel. "Pyrrhic Victory in the 'Tournament of Shadows': Central Asia's Quest for Water Security (1991–2009)." *Asian Security* 6, no. 2 (2010): 121–145.

Chazan, Vigal. "How Viable Is Russia and Iran's 'Sanctions-Evasion' Corridor?" *41 Geopolitical Monitor*. April 6, 2023.

Chotiner, Isaac. "Why Russian Elites Think Putin's War Is Doomed to Fail." *The New Yorker*. May 3, 2023.

Chupilkin, Maxim, Beata Javorcik, and Alexander Plekhanov. "The Eurasian Roundabout: Trade Flows into Russia through the Caucasus and Central Asia." EBRD Working Paper No. 276, European Bank for Reconstruction and Development. February, 2023.

Cohen, Ariel. *Kazakhstan: The Road to Independence – Energy Policy and the Birth of a Nation*. Washington, DC: Johns Hopkins SAIS/Silk Road Studies Program, 2008.

Cooley, Alexander. *Great Games, Local Rules: The New Great Power Contest in Central Asia*. Oxford: Oxford University Press, 2012.

Cooley, Alexander, and John Heathershaw. *Dictators Without Borders: Power and Money in Central Asia*. New Haven, CT: Yale University Press, 2017.

De Waal, Thomas. *The Caucasus: An Introduction*. Oxford: Oxford University Press, 2010.

Dugin, Alexander. *The Foundations of Geopolitics*. Russian Edition. Moscow Arktogia, 1997.

Dugin, Alexander. *Eurasian Mission: An Introduction to Neo-Eurasianism*. London: Arktos, 2014.

Dugin, Alexander. *Putin vs. Putin: Vladimir Putin Viewed from the Right*. London: Arktos, 2014.

Dugin, Alexander. *Last War of the World-Island: The Geopolitics of Contemporary Russia*. London: Arktos, 2015.

Fingar, Thomas (ed.), *The New Great Game: China and South and Central Asia in the Era of Reform*. Stanford, CA: Stanford University Press, 2016.

Fishelson, James. "From the Silk Road to Chevron: The Geopolitics of Oil Pipelines in Central Asia." Geohistory. School of Russian and Asian Studies. Winter, 2007.

Gachechiladze, Revaz. *The New Georgia: Space, Society, Politics*. College Station: Texas A&M University Press, 1995.

Goldman, Marshall I. *What Went Wrong with Perestroika*. New York: W.W. Norton, 1991.

Goldman, Marshall I. *Lost Opportunity: Why Economic Reforms in Russia Have Not Worked*. New York: W.W. Norton, 1996.

Goldman, Marshall I. *The Piratization of Russia: Russian Reform Goes Awry*. New York: Routledge, 2003.

Goldman, Marshall I. *Petrostate: Putin, Power and the New Russia*. New York: Oxford University Press, 2008. (also published as Oilopoly by OneWorld Publications, 2008.)

Gollwitzer, Heinz. *Europe in the Age of Imperialism: 1880–1914*. New York: Harcourt, Brace and World, 1969.

Goltz, Thomas. *Georgia Diary: A Chronicle of War and Political Chaos in the Post-Soviet Caucasus*. Armonk, NY: M.E. Sharpe, 2009.

Graham, Norman A. "The Prospect for Regional Governance of Inland Fisheries in Central Eurasia." Paper prepared for the FAO-MSU Global Conference on Inland Fisheries, January, Rome. Published in Select Proceedings of the Conference in 2016.

Graham, Norman A., and Cody Schulz. "The Political Economy of Economic Diversification in Azerbaijan, Kazakhstan and Russia: Progress and Challenges." The Great Steppe II: Papers presented to the Forum of Humanities of the International Turkic Academy. Astana, Kazakhstan. May, 2017.

Graham, Norman A., Stephen Pueppke, and Talgat Uderbayev. "The Current Status and Future of Central Asia's Fish and Fisheries: Confronting a Wicked Problem." *Water* 9 (August, 2017): 1–16.

Graham, Norman A., Stephen G. Pueppke, Sabir Nurtazin, Talgarbay Konysbayev, Farid Gibadulin, and Meiirli Sailauov. "The Changing Dynamics of Kazakhstan's Fisheries Sector: From the Early Soviet Era to the Twenty-First Century." *Water* 14 (May, 2022): 1–22.

Hale, Henry E. *Patronal Politics: Eurasian Regime Dynamics in Comparative Perspective*. Cambridge: Cambridge University Press, 2014.

Heiser, James D. *"The American Empire Should Be Destroyed" Aleksander Dugin and the Perils of Immantized Eschatology*. Malone, TX: Repristination Press, 2014.

Jones Luong, Pauline, and Erika Weinthal. *Oil Is Not a Curse*. New York: Cambridge University Press, 2010.

Jonson, Lena. *Vladimir Putin and Central Asia: The Shaping of Russian Foreign Policy*. London: I.B. Taurus, 2006.

International Crisis Group. Central Asia: Water and Conflict. ICG Asia Report No. 34. Osh/Brussels. May 30, 2002.

International Crisis Group. The Curse of Cotton: Central Asia's Destructive Monoculture. Asia Report No. 93. Bishkek/Brussels. February 28, 2005.

Karimov, Islam. *Uzbekistan on the Threshold of the Twenty-First Century: Challenges to Stability and Progress*. New York: St. Martin's Press, 1998.

Kazakhstan. *Message of President Nursultan Nazarbayev. Prosperity, Security and Ever Growing Welfare of All the Kazakhstanis*. (7 Long Term Priorities and Agenda for 1998–2000), 1997.

Kazakhstan. Message of President Nursultan Nazarbayev. Invest in Kazakhstan, 2016.

Kazakhstan. Message of President Nursultan Nazarbayev. Third Stage of Modernization in Kazakhstan: 5 Priorities. January 31, 2017.

Kleveman, Lutz. *The New Great Game: Blood and Oil in Central Asia*. London: Atlantic Books, 2003.

Kukeeva, Fatima, and Tolganay Ormysheva. "The Ili River: Is It a 'Casualty' of Kazakhstan's Complicated Relationship with China." Paper presented at the Annual Conference of the Central Eurasian Studies Society, University of Washington, Seattle. October, 2017.

Laruelle, Marlène. *Russian Eurasianism: An Ideology of Empire*. Baltimore, MD: Johns Hopkins University Press, 2008.

Laruelle, Marlène, and Jean Radvanyi. *Understanding Russia: The Challenges of Transformation*. Lanham, MD: Rowman and Littlefield, 2019.
Lavigne, Marie. *The Economics of Transition: From Socialist Economy to Market Economy*. Second Edition. New York: St. Martin's Press, 1999.
Levine, Steve. *The Oil and the Glory: The Pursuit of Empire and Fortune on the Caspian Sea*. New York: Random House, 2007.
Lieven, Dominic. *Empire: The Russian Empire and Its Rivals*. New Haven, CT: Yale University Press, 2000.
Madramootoo, Chandra A., and Victor A. Dukhovny (eds.), *Water and Food Security in Central Asia*. Dordrecht, Netherlands: Springer/NATO Science for Peace and Security Series-C: Environmental Security, 2011.
Marriott, James, and Mika Minio-Paluello. *The Oil Road: Journeys from the Caspian Sea to the City of London*. London: Verso, 2012.
McCauley, Martin. *Khrushchev and the Development of Soviet Agriculture: The Virgin Land Programme 1953–1964*. New York: Holmes & Meier Publishers, Inc., 1976.
Menon, Rajan, Yuri E. Fedorov, and Ghia Nodia (eds.), *Russia, the Caucasus, and Central Asia: The 21st Century Security Environment*. New York: M.E. Sharpe, 1999.
Michel, Casey. "Tajikistan's Rogun Dam Rankles Uzbekistan." The Diplomat. July 21, 2016.
Misiunas, Romuald, and Rein Taagepera, *The Baltic States: Years of Dependence 1940–1990*. London: Hurst and Co., 1993.
Nazarbaev, Nursultan. *The Kazakhstan Way*. London: Stacey International, 2008.
Nikanova, Svetlana, and Tatyana Nazarova. *Aral Yesterday and Now: Problems and Perspectives of Aral's Crisis*. (photoalbum) Almaty, KZ: MFSA, 1997.
OECD. *Review of Agricultural Policies: Kazakhstan*, 2013.
Olcott, Martha Brill. *The Kazakhs*. Stanford, CA: Stanford University, Hoover Institution Press. Second edition, 1995.
Olcott, Martha Brill. *Central Asia's New States: Independence, Foreign Policy and Regional Security*. Washington, DC: U.S. Institute of Peace Press, 1996.
Olcott, Martha Brill. *Central Asia's Second Chance*. Washington, DC: Carnegie Endowment for International Peace, 2005.
Olcott, Martha Brill. *Kazakhstan: Unfulfilled Promise?* Washington, DC: Carnegie Endowment for International Peace, 2010.
Olcott, Martha Brill. *Tajikistan's Difficult Development Path*. Washington, DC: Carnegie Endowment for International Peace, 2012.
Overland, Indra, Heidi Kjaernet, and Andrea Kendall-Taylor (eds.), *Caspian Energy Politics: Azerbaijan, Kazakhstan and Turkmenistan*. London: Routledge, 2010.
Pannier, Bruce. "Turmenistan's Gas Gloom." OilPrice.com. March 11, 2017.
Pannier, Bruce, and Edige Magauin. "Kazakhstan: China Discusses Future of Irtysh River." RadioFreeEurope/RadioLiberty Archive. May 9, 1999. https://www.rferl,.org/a/1091429.html.
Pavlov, L. "Cotton in Russia." *The Russian Review* 1 (May, 1916): 237–240.
Peimani, Hooman. *Failed Transition, Bleak Future? War and Instability in Central Asia and the Caucasus*. Praeger, Westport, CT: Praeger, 2002.
Peyrouse, Sebastien. "The Hydroelectric Sector in Central Asia and the Growing Role of China." *China and Eurasia Forum Quarterly* 5, no. 2 (2007): 131–148.
Plokhy, Serhii. *The Last Empire: The Final Days of the Soviet Union*. New York: Basic Books, 2014.

Pomfret, Richard. *Kazakhstan's 2030 Strategy: Goals, Instruments and Performance*. Paper presented at the American Economic Association Annual Conference. January 4, 2014.

Porter, Michael E. *Kazakhstan's Competitiveness: Roadmap Towards a Diversified Economy*. Boston, MA: Harvard Business School, Institute for Strategy and Competitiveness, 2005.

Pueppke, Steven J., Sabir T. Nurtazin, Norman A. Graham, and Jiaguo Qi. "Central Asia's Ili River Ecosystem as a Wicked Problem: Unraveling Complex Interrelationships at the Interface of Water, Energy, and Food." *Water* 10 (April, 2018): 1–21.

Putz, Catherine. "Uzbekistan's Changing Rogun Tone." *The Diplomat*. July 10, 2017.

Putz, Catherine. "China's Imports of Turkmen Gas Decline." *The Diplomat*. December 27, 2017.

Robbins, Christopher. *Apples are from Kazakhstan: The Land that Disappeared*. New York: Atlas and Co., 2008.

Rodrik, D., A. Subramanian, and F. Trebbi. "Institutions Rule: The Primacy of Institutions Over Geography and Integration in Economic Development." *Journal of Economic Growth* 9, no. 2 (2004): 131–165. http://dx.doi.org/10.1023/b:joeg.0000031425.72248.85

Ross, M. *The Oil Curse: How Petroleum Wealth Shapes the Development of Nations*. Princeton, NJ: Princeton University Press, 2012.

Roy, Olivier. *The New Central Asia: The Creation of Nations*. New York: New York University Press, 2000.

Sachs, J., and A. Warner. Natural Resource Abundance and Economic Growth. NBER Working Paper Series, (Working Paper 5398). 1995. http://dx.doi.org/10.3386/w5398

Sarymbetova, Aygerim. "Kazakhstan to Start Gas Deliveries to China." *Caspian News*. October 10, 2017.

Starr, S. Frederick, and Svante E. Cornell (ed.), *Uzbekistan's New Face*. Lanham, MD: Rowman and Littlefield, 2018.

Stucki, Virpi, Kai Wegerich, Muhammad Mizanur Rahaman, and Olli Varis (eds.), *Water and Security in Central Asia: Solving a Rubik's Cube*. London: Routledge, 2014.

Tekin, Ali, and Paul Andrew Williams. *Geo-Politics of the Euro-Asia Energy Nexus: The European Union, Russia and Turkey*. New York: Palgrave Macmillan, 2011.

Thorpe, Andy, et al. "The Collapse of the Fisheries Sector in Kyrgyzstan: An Analysis of Its Roots and Its Prospects for Revival." *Communist and Post-Communist Studies* 42 (2009): 141–163.

Trenin, Dmitri. *The End of Eurasia: Russia on the Border Between Geopolitics and Globalization*. Washington, DC: Carnegie Endowment for International Peace, 2002.

Utegenova, Ailuna R. *Kazakhstan's 2030 Development Strategy: Significance and Results*. 2010. https://ifsh.de/file-CORE/documents/yearbook/english/10/Utegenova-en.pdf

Westerman, Frank. *Engineers of the Soul: The Grandiose Propaganda of Stalin's Russia*. New York: The Overlook Press, 2002.

Wheeler, William. *Environment and Post-Soviet Transformation in Kazakhstan's Aral Sea Region*. London: UCL Press, 2021.

Whitman, John. "Turkestan Cotton in Imperial Russia." *The American Slavic and East European Review* 15, no. 2 (April, 1956): 190–205.

Wittfogel, Karl A. *Oriental Despotism: A Comparative Study of Total Power*. New Haven, CT: Yale University Press/New York: Vintage Books, 1957, 1981.

World Bank. *Kazakhstan: Transition of the State*. Washington, DC, 1997.

World Bank. *Europe and Central Asia Region. Water Energy Nexus in Central Asia: Improving Regional Cooperation in the Syr Darya Basin*. Washington, DC, 2004.

World Bank. *Assessment Studies for Proposed Rogun Hydropower Project in Tajikistan.* Washington, DC. September 1, 2014.

Zhanbosinova, Albina, Fatima Kukeyeva, Tolganay Ormysheva, and Norman A. Graham. "Irtysh Whirls: The Geopolitics of River Rights Between Kazakhstan, China and Russia." CERES Occasional Papers. Center for European, Russian and Eurasian Studies. Michigan State University, July, 2019.

Zinzani, Andrea. *The Logics of Water Policies in Central Asia: The IWRM Implementation in Uzbekistan and Kazakhstan.* Zurich, Switzerland: LIT Verlag, 2014.

Zonn, Igor S., and Audrey G. Kostianoy (eds.), *The Turkmen Lake Altyn Asyr and Water Resources in Turkmenistan.* Berlin: Springer-Verlag, 2014.

6

CHINA'S BELT AND ROAD INITIATIVE

Challenges and Opportunities for Central Asia in the Energy-Water-Food Nexus

The People's Republic of China (PRC) is a critical actor for the energy-water-food nexus in Central Asia. In return, Central Asia also plays a decisive role in China's strategic calculus and ambitions for regional and global power projection.[1] In 2013, the Chinese Government announced two ambitious multi-billion dollar regional integration initiatives, the Silk Road Economic Belt and the twenty-first century Maritime Silk Road, which are jointly referred to as the Belt and Road Initiative (BRI).[2] These comprehensive infrastructure development projects of an unprecedented scale are projected to span over a broad geographical space including Eurasia and East Africa while also connecting China with Europe, the Indian Ocean, and the Persian Gulf.[3] At the regional level, due to their geopolitical and economic significance, the Central Asian states are also integrated as partners in the BRI. In this context, due to Central Asia's potential in terms of energy reserves and development, its strategic location on major energy supply routes, and its proximity to the geopolitically sensitive Xinjiang Uygur Autonomous Region, it is crucial to explore the energy dimension of the BRI in Central Asia. Moreover, the pressing environmental problems and water scarcity aggravated by climate change in this energy-rich, yet water-poor region, brings rather neglected dimensions of water and food security and sustainability into the limelight as well.

At the global level, eroding US hegemony and rise of China's economic power intensifies US-China competition. Hence, BRI emerges as a significant puzzle piece in the escalating competition and tensions, as well as the US efforts to counter China's increasing regional and global influence.[4] In *Exit from Hegemony*, Alexander Cooley and Daniel Nexon posit that the liberal international order is experiencing a fundamental transformation. Russia and China have emerged as revisionist powers challenging the existing hierarchy by forming alternative regional institutions and promoting counter-norms.[5] Since energy security is vital for steady economic

DOI: 10.4324/9781003373735-6

growth, China also emerged as a decisive player in global energy dynamics as the world's second-largest economy and the leading energy consumer since 2010.[6]

In crafting its strategic positioning in Central Asia and rapidly expanding its economic sphere of influence, China has been pursuing a combination of hard and soft power tactics, including military power, economic means, diplomacy, and political leveraging to advance its energy interests and security. In relation to our triangular theoretical framework of Security, Energy, and Environment, our analysis reveals that while ensuring security and access to energy resources constitute China's priority, environmental concerns and water and food security related issues received considerably less attention. On the contrary, there are significant concerns that through BRI China will be shifting some of its most polluting and water consuming industries to Central Asia. Moreover, increasing agricultural sales to China from this region, while on the one hand boosts the economies of Central Asian countries, on the other hand further strains their dwindling water resources.

This chapter examines the challenges and opportunities for China's interaction with the Central Asian republics for energy, water, and food security. It is also noteworthy that there are significant variations among the Central Asian states. We posit that under the framework of the BRI, China has been combining various types of power in obtaining collaboration and compliance of partner states. As China strives to advance its position in global politics and economy and to ensure its energy security, the Central Asian states also try to benefit from their complex regional ties with external actors at a time of persisting Russian political influence and increasing Chinese economic engagement at the regional level and intensifying US-China competition in the global context. At the national and local level, the Central Asian states, while trying to enhance their economic benefits, will also need to cope with rising environmental challenges with varying degrees of success.

China's Quest for Energy Security and Implications for Central Asia

Chinese scholar Zha Daoijiong argues that "A country has meaningful energy security only when its management of balance in energy supply and demand serves the purpose of developing its economy and society well."[7] Moreover, the need to confront global warming has also become increasingly salient for energy security debates. Hence, energy security has emerged as a multidimensional concern giving way to a complex web of relations for all relevant actors.

After the proclamation of the People's Republic of China in 1949, energy became a priority area in the Communist Party's agenda. During the 1949–1978 period, while targeting industrial development, PRC strived to be self-sufficient and self-reliant pursuing a coal-based and oil-complemented energy strategy. However, following Deng Xiaoping's 1978 economic reforms, there was a major transformation in energy policies. China's rapid economic growth also led to a dramatic increase in its energy needs. Thus, China's status changed from a net oil exporter to a net oil importer in 1993, constituting a critical milestone in its energy politics.[8] This shift compelled China to intensify its efforts for diversification and for the pursuit of a supply-oriented concept of energy security.

In July 2006, the former Chinese President Hu Jintao introduced the "New Energy Security Concept," marking Beijing's embracement of a global understanding of energy security. Currently, China pursues a comprehensive approach coined by Xi Jinping, targeting revolutions in energy consumption, supply, technology, and systems while also advancing international cooperation.[9] Accordingly, China signed numerous bilateral and multilateral energy agreements and engaged in collaborative schemes for international energy projects. The primary goals have been to enhance infrastructure, to diversify suppliers, and to mitigate the negative impact of energy choices on climate change. China has also assumed a more salient role in global energy governance. Ultimately, as Zhao argues in *The Economics and Politics of China's Energy Security Transition,* despite some setbacks and limitations, gradually China's "energy policy was shifted from plan-dominated to market-dominated, from self-reliance to interdependence, from single energy development to diversified energy security guarantee."[10]

Deng Xiaoping's reform policies and the opening of 1978 fueled China's economic development and growth over the past four decades. Despite the detrimental impact of the Covid-19 pandemic on its economy, China emerged as the world's second-largest economy. Consequently, China is a decisive player in the global energy scene with a rapidly increasing demand for energy resources. China has ranked as the world's largest energy consumer accounting for 24% of global energy consumption and driving 34% of global energy demand growth in 2018. Moreover, BP projects that China will remain the leading energy consumer by a long stretch into 2040.[11]

Coal dominates China's energy mix with a share of 58%. However, environmental and air quality concerns compel China to seek cleaner alternatives. Hence, China's Blue Sky 2018–2020 policy is aimed at mitigating air pollution and achieving a decrease in coal-dependency.[12] As a result of this shift in policy direction for more environmentally friendly alternatives, the share of coal in energy consumption in 2018 decreased to a historical low.[13] After coal, oil accounts for 20% share in the current energy mix and natural gas constitutes 7.4%. Renewables and nuclear energy have a relatively more limited share of 4.4% and 2%, respectively. China is the largest importer of coal, oil, and natural gas in the global energy markets. For instance, the Chinese import dependency on oil increased to 72.5% in 2019, reaching the highest level in the past fifty years. Its import dependency on natural gas also increased to 40.6% in this period.[14] In 2021, China imported $208B in crude petroleum, ranking as the 1st largest importer of crude petroleum in the world. During the same year, crude petroleum was the most imported product in China. Chinese imports of crude petroleum are supplied primarily by: Saudi Arabia ($38.3B), Russia ($35.4B), Iraq ($23.1B), Oman ($19.7B), and Angola ($19.6B).[15] In 2021,

China imported a total of 156 bcm of natural gas (including LNG). 56% of piped gas imports were supplied by Turkmenistan (31 bcm), 18% by Russia (10 bcm), and 11% by Kazakhstan.[16] In this context, particularly for natural gas, the Central Asian suppliers and supply routes are highly salient.

China views the US dominance in the Middle East, Southeast Asia, and East Asia regions as a critical threat to its supply routes. Since approximately 80% of the imports arrive via the South China Sea, including natural gas and crude oil imports, this region is vital for Chinese interests. The South China Sea and its potential energy reserves cause increasing tensions. The projections regarding offshore gas and oil fields have even led to labeling of this region as the "new Persian Gulf" in some reports.[17] China's sovereignty claims over the South China Sea and legal disputes among the riparian states further complicate the picture. Bypassing the Malacca Strait and finding alternative routes has emerged as a critical challenge for China. Moreover, there are significant concerns regarding the prospects of China's domestic demand outpacing its supplies and causing an energy crisis. The Chinese government strives to mitigate these challenges by facilitating alternatives through the BRI. Hence, China has been targeting the diversification of its energy suppliers with the BRI and to create supply route alternatives beyond the military and political outreach of the United States.

In this context, it is significant to highlight that China's policies and projects in different regions are not solely driven by energy security concerns. Moreover, energy security should be viewed comprehensively as driven by geopolitical goals, economic growth, demographic trends and societal dynamics, and environmental concerns. Ultimately, all these factors have a direct or indirect impact on the Chinese energy security concerns and strategies and China's interaction with the Central Asian states.

The BRI project targets regional cooperation with a vast geographic scope including partner countries extending from Central Asia to Africa. The current number of participant countries increased from 65 in 2017 to 147 in 2022.[18] Five key areas of cooperation between China and the partner countries are: (1) policy coordination, (2) facilities connectivity, (3) unimpeded trade, (4) financial integration, and (5) people-to-people bonds.[19] Through this project, China aims to achieve new markets for deploying its overcapacities, construct adequate infrastructure within and between states, consolidate its diplomatic ties, and advance its energy security interests.

China authorities have been utilizing a combination of attraction and coercion strategies regarding the BRI. In terms of attractions, China provided debt and extended credit to some partner countries and enabled some other countries (i.e. Kingdom of Tonga) to defer their debts as they joined the BRI as signatories.[20] China also appealed to partner countries with the promise of making the BRI a green, clean, and sustainable initiative[21] and became the largest contributor to renewable energy systems in 2018.[22] As for coercive strategies, China increased its military presence by expanding its naval power particularly in the South China

Sea.[23] In terms of energy relations, China has a much more powerful stance in comparison to individual countries along the BRI. Except for Russia, in every other case, China benefits from higher relative gains and develops asymmetric patterns of interdependence.[24] Consequently, BRI equips China with an extensive policy toolbox enabling Beijing to transform into a more influential player in the regional and global context.

China's Regional Cooperation and Energy Interactions with Central Asian Republics

After the collapse of the Soviet Union in 1991, the Central Asian republics (Kazakhstan, Kyrgyzstan, Tajikistan, Turkmenistan, and Uzbekistan) have become increasingly more important for Beijing. As these states opened up to the global political and economic environment, this region which was once referred to as "Russia's backyard" came under Chinese economic sway, even though strong Russian influence persists.[25] The foundation of the Shanghai Cooperation Organization (2001) established the framework for deepening partnerships between China and the Central Asian republics, which is becoming more entrenched with the Belt and Road Initiative. We posit that Central Asia is significant for Beijing's strategic calculus for three main reasons: (i) for its critical location on energy supply and trade routes; (ii) attractive hydrocarbon resources and raw materials; and (iii) its proximity to the Xinjiang Uyghur Autonomous Region (XUAR). It is also noteworthy that each Central Asian state has its own level of integration to the BRI and has varying degrees of significance for energy security. Hence, as also stressed by Laurelle

> not all Central Asian states are involved to the same degree: Kazakhstan is at the forefront, followed by Tajikistan and Kyrgyzstan. Uzbekistan is somewhat on the sidelines… For the most part, Turkmenistan's gas trade is already dominated by China, but this is a specific sector that has few ties with BRI infrastructure programs.[26]

To enhance international connectivity, two out of the six economic corridors of the BRI passes through Central Asia.[27] The New Eurasian Land Bridge, which starts from Yiwu in eastern China, extends all the way to Germany in the European Union. The China-Central Asia-Western Asia corridor connects Xinjiang to the Port of Piraeus in Greece after crossing the Middle East. Besides these economic corridors, China and Central Asian republics have been collaborating in numerous energy, infrastructure, and railroad projects. So far, three railroad projects have been constructed: The Pap-Angren electrified railway line to Uzbekistan; Uzen-Bereket-Gorgan which goes through Kazakhstan, Turkmenistan, and Iran; and Horgos dry port which connects China to Kazakhstan.[28] In the energy realm, numerous Chinese companies are actively engaged in this region: China National Petroleum Corporation (CNPC), China National Offshore Oil Corporation (CNOOC), China

Petroleum and Chemical Corporation (SINOPEC), and PetroChina have been partnering with local companies for exploration and extraction of oil and natural gas.[29] Moreover, two essential pipelines connect Central Asia to China. The Central Asia-China gas pipeline transports natural gas through Turkmenistan, Uzbekistan, and Kazakhstan and reaches Horgos which is situated in Xinjiang. It then connects to China's west-east pipeline and transports natural gas from Central Asia to Hong Kong. This pipeline has been carrying 15% of natural gas needs to Chinese borders.[30] Moreover, the Kazakhstan-China oil pipeline enables Chinese companies to control 25% of Kazakhstan's total oil production.[31] While Kazakhstan's share in China's oil import market is relatively small, prospective projects turn Astana into an important partner. Under the framework of the BRI, China projects to increase the capacity of these pipelines and expand energy cooperation with Central Asian states.

Central Asia is also adjacent to Xinjiang, which is a highly significant and sensitive region for Chinese domestic politics. The presence of the main oil resources of China combined with ethnic and political tensions presents a complex setting. Xinjiang has one-fifth of China's total fossil fuel reserves, and overall holds the largest national reserves of coal, natural gas, and oil.[32] Moreover, in addition to discoveries of new fields,[33] this region has a high potential for renewable energy resources such as wind and solar. The Chinese government stressed the significance of the region for its energy security in the 13th Five Year Plan for Economic and Social Development of the PRC (2016–2020) and pledged to make XUAR the "core region" for the Silk Road Economic Belt.[34] However, ethnic strife and oppressive government policies such as continuous surveillance and patrols against the Uyghur population cast a shadow over the region's economic prospects. Given the geopolitical and strategic salience of the Xinjiang region, Central Asian republics are also affected by these problems since Kazakhstan, Kyrgyzstan, and Tajikistan share borders with this region. Moreover, while accommodating substantial portion of the Uyghur population within their countries, they also provide shelter for those who flee China resulting in bilateral tensions.

When China turned into a net oil importer in 1993, energy started to play a more decisive role in Beijing's calculus regarding Central Asia.[35] China aims to diversify its suppliers and take part in the exploration and development of new fields through its deepening links with Central Asia. In this context, Kazakhstan, Turkmenistan, and Uzbekistan have a more prominent share in terms of China's energy imports from Central Asia. Table 6.1 reflects the variation among different Central Asian republics regarding their exports in mineral fuels, mineral oils, and products of their distillation. While Kazakhstan has the lions share, Kyrgyzstan and Tajikistan have a negligible amount.

Among the five republics with varying reserves of energy resources, Kazakhstan, Turkmenistan, and Uzbekistan are particularly appealing for the Chinese investors and the international energy markets. According to the IEA, Kazakhstan is among the top producers of crude oil in Europe and Eurasia. It is also a major

TABLE 6.1 Energy Resource Imports of China from the Central Asian Republics[36]

	Energy Imports and Export Data: Mineral Fuels, Mineral Oils and Products of Their Distillation	*Value in 2020*	*Value in 2021*
Kazakhstan	China's imports from Kazakhstan	1,421,877	2,639,070
	Kazakhstan's exports to world	27,329,211	34,820,911
Kyrgyzstan	China's imports from Kyrgyzstan	644	5
	Kyrgyzstan's exports to world	81,723	99,988
Uzbekistan	China's imports from Uzbekistan	63,948	15,752
	Uzbekistan's exports to world	642,287	883,391
Tajikistan	China's imports from Tajikistan	0	0
	Tajikistan's exports to world	55,654	96,060
Turkmenistan	China's imports from Turkmenistan	741,564	459,400
	Turkmenistan's exports to world	1,525,331	1,847,186

(*Unit*: US Dollar thousand).

producer of coal and natural gas with increasing renewable energy sources. Uzbekistan is self-sufficient owing to its natural gas resources and China spent $585 million to support oil and gas projects in this country.[37] Turkmenistan holds the fourth-largest global gas reserves and it is the biggest natural gas exporter to China. For Turkmenistan, as its top gas export destination, energy sales to China constituted 61% of its gas exports in 2018.[38] Although Kyrgyzstan and Tajikistan have much more modest energy resources, they are also significant for China's energy security due to their strategic location on critical supply routes which constitute a bridge between China and the Persian Gulf with its rich energy and gas reserves.

China's Power Projection Strategies in Central Asia

In pursuit of energy security, the power projection strategies of China in Central Asia can be examined in four categories, namely military, economic, political/diplomatic, and technological.

In terms of military and hard security strategies, China has been conducting joint counterterrorism operations with the Central Asian republics. These operations mainly target terrorism, separatism, and religious extremism. The Shanghai Cooperation Organization (SCO) has become the primary security organization for Central Asia, surpassing the Collective Security Treaty Organization (CSTO). It is noteworthy that the SCO also established anti-terrorism networks within member states such as the Regional Anti-Terrorist Structure (RATS). In addition to the broader initiatives under the umbrella of the SCO, China also has some specific bilateral collaboration schemes with individual countries. For instance, China established a military base in Tajikistan. This base is located in proximity to the Xinjiang border and is established under the auspices of protecting the Wakhan corridor in Afghanistan, which is an important station for the BRI.[39] In facing the conflicts between the Han Chinese and Uyghur Muslims, Chinese military presence is quite influential in Xinjiang and China also strives to control its immediate borders. In addition to alleviating its security concerns, China aims to protect its pipelines and ongoing energy projects by obtaining political stability in this region and its vicinity.

Economically, China currently dethroned Russia as the biggest trading partner of Central Asia. It has become the largest trading partner of Kazakhstan and Turkmenistan, the second-largest trading partner of Uzbekistan and Kyrgyzstan, and the third-largest trading partner of Tajikistan.[40] Intensifying trade ties creates a more conducive setting for enhancing energy supply security and boosting regional economic development. China has also tried to form a free trade area with the Central Asian states; however, Russia has curtailed this initiative. While the existence of the Eurasian Economic Union undermined China's attempts, with the BRI, China has facilitated effective mechanisms for more extensive penetration into the Central Asian markets. Utilizing the AIIB funds and Silk Road Fund, China has been lending substantial amounts of loans and its state-owned enterprises and firms have been investing in infrastructure, transportation, and energy networks in the region. President Xi Jinping made official visits to every Central Asian state and promised to invest 40 billion dollars in the region's infrastructure projects through the Silk Road Infrastructure Fund. However, Beijing's lending and assistance packages frequently entail some implicit conditions. For example, debtor countries are expected to recognize the "One China Policy" and to acknowledge the PRC as the legitimate government of all China. Moreover, Chinese investments might include some strings necessitating compliance with China's policy toward the Uyghur population.[41]

The lending structure might also provide asymmetric benefits for the Chinese lenders. To exemplify the broader economic gains out of debt-giving, two cases in Kyrgyzstan and Turkmenistan are particularly illustrative. In 2006, China offered to provide Kyrgyzstan 1.2 billion dollars in preferential credits to build a railway in exchange for access to mineral resources.[42] In 2009, in return for benefiting from the South Yolotan, one of the largest natural gas fields in the world, China extended a 3 billion dollar loan to Turkmenistan. This loan is part of a 30-year agreement, which provides China with approximately 40 billion cubic meters of gas per year.[43] Supported by their central government, Chinese firms also benefit from concessional loans to Central Asia. These loans regarding different projects ranging from infrastructure to technical assistance stipulate that no less than half of the materials, equipment, technology, and services provided under the contract should be acquired from China.[44] This strategy profits Chinese enterprises, enables them to expand abroad, and more easily obtain overseas contracts and resources, reinforcing China's deepening engagement in Central Asian markets.

In the political and diplomatic realm, in line with the "Good Neighbor Policy" spearheaded by Hu Jintao, Xi Jinping has pursued similar policies to build amiable political and diplomatic ties with the Central Asian states. In 2013, Xi Jinping made the first announcement regarding the Silk Road Economic Belt at the Nazarbayev University in Kazakhstan. There have been numerous high-level state visits to Central Asia highlighting the significance of this region for China. According to a report published in 2019, between 2000 and 2017, there have been a total of 722 visits between Chinese and Central Asian government officials.[45] During all of these visits, both sides stressed the importance of their strategic partnership and friendly relations, as well as their commitment for strengthening their ties under the frameworks of SCO and BRI.

A critical field of cooperation particularly desired by the Central Asian republics is technology. Central Asian governments view technological development as a decisive factor for their advancement. In this context, one of the promises made by the Communist Party regarding the BRI was to advance the technological capabilities of the partner countries by integrating them into the Digital Silk Road. The Digital Silk Road aims to create an urban transformation towards smart cities utilizing the latest technologies. Chinese firms such as Huawei are exporting their technologies to the Central Asian republics under this framework. Chinese energy and infrastructure-oriented companies also offer training and expertise regarding the recent technological advances in this field to Central Asian countries in a number of projects. On the more alarming side, there are also concerns about China's increasing digital dominance over the region with its surveillance systems.[46]

Challenges and Constraints in the Regional and Local Context

In terms of challenges and constraints, expansion of Chinese influence in Central Asia intensifies its competition with other major regional players such as Russia and India. Central Asian states are already part of many agreements and organizations with the Russian Federation such as the Eurasian Economic Union (EEU). Although Vladimir Putin tries to strike a delicate balance in Russia's complex relations with China in countering Western influence and institutions, Russian oligarchs and conglomerates have a more antagonistic stance towards the increasing engagement of Chinese firms in Central Asia.[47] In formulating a "cautious, complex and pragmatic" response towards the BRI, the Indian government is skeptical towards Beijing regarding the implications of the BRI as well. In particular, India criticizes Chinese pipeline initiatives, which are undermining its existing projects such as the Turkmenistan-Afghanistan-Pakistan-India (TAPI) pipeline.[48] The United States also views expansion of Beijing's power and influence fueled by BRI as a major challenge in its intensifying regional and global contestation with China.[49]

At the local and national level, Chinese migrants who work in projects within the Central Asian states are in some instances met with skepticism and even hostility. Media and populist politics due to the significance of China's demographic power aggravate these fears. For instance, one local newspaper voiced these concerns bluntly stating that Kyrgyz people were deeply suspicious that "the enormous difference in the size of the two countries and peoples would mean that the Kyrgyz would soon drown in a Chinese sea."[50] Rising Sinophobia and mistrust strain the people-to-people connections between the regions. Moreover, there are increasing concerns regarding the adverse implications of Chinese debts and economic involvement in the region. According to a 2018 report, Kyrgyzstan and Tajikistan are among the eight countries that are predicted to experience a "debt distress" due to BRI lending.[51] China is repeatedly accused of pursuing "debt-trap diplomacy" and its lack of transparency in the projects has been an issue of contention with partner countries.

Hence, despite the vast opportunities that BRI provides for collaborative schemes particularly in the realm of energy projects and security, it also presents significant challenges for China's regional and global competitors, as well as its Central Asian partners. The water and food dimension of China's increasing engagement in Central Asia will also have vital implications for the future dynamics.

How Will BRI Impact Central Asia's Water-Food Nexus and Sustainability ?

For Central Asia, while China's BRI emerges as the most extensive infrastructure initiative in generations, reminiscent of Soviet era mega dams and cotton monoculture, BRI also deepens the problems of water sharing and pollution further endangering precarious water resources across Central Asia. Since development projects envisioned by BRI will further strain already dysfunctional and disjointed water management systems, Davies and Matthews argue that water constitutes "possibly the most serious problem along the [BRI's] route."[52]

After their independence from the Soviet Union in 1991, the Central Asian republics had flaring disputes regarding once common water resources managed by Moscow. For instance, in 2012, President Islam Karimov of Uzbekistan had an inflammatory rhetoric against neighboring Kyrgyzstan and Tajikistan threatening that their efforts to build hydroelectric power stations as upstream riparians could spark war.[53] While at the national level, the change of leadership in Uzbekistan in 2016 has subsided some of the most heated disputes, at the regional level, five Central Asian Republics displayed very limited coordination in dealing with aggravating water shortages exacerbated by increasing populations and climate change.

Can China play a more unifying and constructive role regarding water issues to prevent instability along its western borders within the framework of BRI? There seem to be substantial challenges in this context. At the local and national level, lack of coordination among BRI contractors, opaque deals conducted by Chinese companies and lack of transparency, Chinese disregard for public opinion in host countries, and rising skepticism of the affected populations emerge as some of the most pressing problems. Moreover, unlike energy security, water management and water security does not constitute a priority area for BRI. This rather negligent stance is reflected by the fact that "Water resources and their management currently do not appear to be specifically referenced in any policy documents or proposals relating to the BRI; and the initiative does not have any associated transboundary water cooperation frameworks."[54]

China has a janus-faced approach towards dealing with its increasing global environmental impact with a widening rift between its new domestic environmental priorities and its policies abroad. Domestically, China is aiming to move towards a path of "ecological modernization," particularly in sectors experiencing Western market pressures and in recently expanding fields such as renewable energy, by expressing its commitment to build its own "ecological civilization."[55] However, in terms of its international development projects involving non-Western countries from which China acquires a significant share of its natural resources, the domestic greening of China's economy does not necessarily impact Chinese planning of the BRI. In particular, an analysis of the existing documents and statements associated with this initiative demonstrates scant attention to strategic environmental assessment (SEA) or environmental impact assessment (EIA).[56] Moreover, as China intensifies its efforts to green its own economy, Beijing projects to export some of its polluting industries along the trajectory of the BRI to Central Asia. For instance, cement production is among the most polluting industries that raises concerns. Chinese finance boosted Tajikistan's cement output to increase tenfold within a period of five years by 2020. However, while experiencing this meteoric rise in production levels, neither Tajikistan's leaders took any measures to regulate the industry's detrimental environmental impact, nor the Chinese investors engaged local stakeholders and addressed their concerns.[57]

On the food security front, the operations of China's small and large-scale agricultural enterprises and land acquisitions in Central Asia constitute a significant food security related dimension. In a comprehensive study zooming on Chinese farms in Tajikistan, Hofman argues that

> Chinese land investments in Tajikistan are shaped by Chinese dynamics of agrarian change and are contingent upon country-specific conditions. Rather than state-led endeavors as is often assumed, the main Chinese actors in Tajik agriculture are capitalist yet partially state-embedded enterprises driven by profit-oriented goals.[58]

In this context, large Chinese enterprises particularly target the segments of the domestic market, which have been underdeveloped during the process of post-socialist agrarian change.

As China emerges as one of largest global food importers, the Central Asian governments are also eager to expand their market share for increasing their non-energy based export revenues and generating employment opportunities. Moreover, boosting agricultural production could also benefit local populations. However, the tradeoff is that the expansion of agriculture and extensive water utilization for irrigation aggravates the pressure on already scarce water supplies.

One critical issue is whether China will share its expertise about efficient water usage and assist in repairing the decrepit Soviet water infrastructure? For instance, in the Ferghana Valley, the water canals are in dire need for repairs and there is a drastic amount of waste. This situation could also create some opportunities, since "China's Xinjiang province, which will be closely linked to Central Asia through the BRI, possesses advanced knowledge and technologies in terms of improving irrigation efficiency and tackling issues of salinization."[59] However, currently, the specifics of agricultural initiatives and projects remain largely unclear. Moreover, since many Chinese companies deploy their own workers, the anticipated local employment benefits may remain more limited. The most critical risk remains as the degradation and overexploitation of the regions dwindling water resources.

Conclusion

China has achieved high levels of sustained economic growth since the last four decades. In this context, energy security is a major challenge for Chinese policy-makers with multifaceted implications in the social, economic, and environmental realm. On the one hand, China strives to ensure a constant flow of energy supplies at affordable prices to meet its rapidly rising domestic energy demand. On the other hand, there are also increasing environmental concerns due to China's coal-dominated energy mix. Moreover, the intrusion of US influence in its neighboring regions as well as competition from other regional actors such as Russia and India present additional challenges for China.

In response to these challenges, particularly after 2013 with Xi Jinping's presidency and the launching of BRI, China has been pursuing a mix of hard and soft power strategies in Central Asia to ensure its energy security, utilizing a wide array of policies with military, economic, diplomatic/political, and technological dimensions. While we acknowledge that energy security has not been the only driver of the Chinese government in pursuing these strategies, energy interests have played an instrumental role for China as a pressing priority.

For the Central Asian states, China has also emerged as an increasingly significant actor in the economic sphere resulting in rapidly expanding ties. In crafting their strategies towards China, Central Asian states should comprehensively assess what Chinese infrastructure projects and bilateral agreements will entail. While conducting this assessment, they should also factor in their own social, political, and economic dynamics and potential vulnerabilities. The lack of adequate transparency in Chinese initiatives remains a critical concern. The Chinese government's treatment of the Uyghur population and the lack of transparency regarding data and documentation for some of the infrastructure projects raise suspicions against China. BRI projects facilitated China to deeply integrate into the Central Asian politics and economy by vastly expanding the existing ties. However, there are also several challenges to be tackled. There is the imminent possibility of a backlash from the local populations for implementing a "debt-trap diplomacy" and for over-exploiting the capacities of partner countries. Moreover, major regional powers such as Russia and India, as well as the United States, are also highly skeptical about China's expanding influence in this region.

On the overall environmental impact of the BRI, the United Nations Economic and Social Commission for Asia and the Pacific (UNESCAP) report in 2021 warns that

> The Belt and Road Initiative, similar to any large-scale development programme, poses environmental risks and can have adverse effects on the welfare of people if implemented without sufficient regard for sustainability and climate impacts. In addition, the distribution of development impacts can be uneven.[60]

The potential benefits from BRI investments and projects should be balanced against the potential negative effects on the environment and society to provide more inclusive and sustainable development. Another critical aspect is the necessity for a more equitable distribution of benefits among project partners, across geographic areas and among different groups in society.

In relation to our triangular theoretical framework of Security, Energy and Environment, our analysis reveals that in its relations with Central Asia, China and its BRI often prioritizes security and energy dimension at the expense of environmental concerns and sustainability. These competing trends indicate that Central Asia with rich energy reserves will maintain its significance not only for ensuring energy security but also as a critical puzzle piece in the strategic competition in a transforming regional and global order.

Notes

1 Central Asia in this study refers to the five republics that are situated in the central part of Eurasia: Kazakhstan, Kyrgyzstan, Tajikistan, Turkmenistan, and Uzbekistan.
2 This project is also referred to as "One Belt, One Road" (OBOR) or the "New Silk Road." Initially, the initiative was referred to as two separate projects, then as the 'One Belt, One Road' initiative, then finally as the Belt and Road Initiative.
3 Organization for Economic Cooperation and Development (OECD), *China's Belt and Road Initiative in the Global Trade, Investment and Finance Landscape* (2018). https://www.oecd.org/finance/Chinas-Belt-and-Road-Initiative-in-the-global-trade-investment-and-finance-landscape.pdf
4 E. Ashbee, *Countering China: US Responses to the Belt and Road Initiative* (Boulder, CO: Lynne Rienner Publishers, 2023). https://doi.org/10.1515/9781685859442
5 A. Cooley and Daniel H. Nexon. *Exit from Hegemony: The Unraveling of the American Global Order* (New York: Oxford University Press, 2020).
6 World Bank. (n.d.). *The World Bank in China: Overview*. Retrieved from https://www.worldbank.org/en/country/china/overview; Financial Times. *China Now World's Biggest Energy User*, July 19, 2010. Retrieved from https://www.ft.com/content/937fdd2c-934b-11df-bb9a-00144feab49a .
7 Daojiong, Z. "China's Energy Security: Domestic and International Issues," *Survival* 48, no. 1 (2006): pp. 179–190. doi:10.1080/00396330600594322
8 Ibid.

9 CECEP. "Xi Jinping: Promote China's Energy Production and Consumption Revolution Vigorously," June 27, 2014. Retrieved from http://www.cecep.cn/g3621/s12533/t37788.aspx
10 H. Zhao. *The Economics and Politics of China's Energy Security Transition* (Cambridge, MA: Academic Press, 2018).
11 "As Renewables Soar, BP Sees China Hitting Brakes on Energy Growth." *Reuters*, February 14, 2019. Retrieved from https://www.reuters.com/article/us-bp-energy/as-renewables-soar-bp-sees-china-hitting-brakes-on-energy-growth-idUSKCN1Q3244
12 "Three-Year Action Plan Aims for Blue Skies." *China Daily*, June 19, 2018. Retrieved from https://www.chinadaily.com.cn/a/201806/19/WS5b287383a310010f8f59d939.html
13 "China's Energy Market in 2018." *BP Statistical Review* (2019). https://www.bp.com/content/dam/bp/business-sites/en/global/corporate/pdfs/energy-economics/statistical-review/bp-stats-review-2019-china-insights.pdf
14 International Energy Agency (IEA), "Oil, Gas and Coal Import Dependency in China, 2007–2019." (2022). Retrieved from https://www.iea.org/data-and-statistics/charts/il-gas-and-coal-import-dependency-in-china-2007-2019
15 The Observatory of Economic Complexity (OEC), *OEC Country Profiles: China*, Crude Petroleum (2023). https://oec.world/en/profile/bilateral-product/crude-petroleum/reporter/chn
16 Enerdata, "China-Central Asia Natural Gas Pipeline Supplied 43 bcm to China." https://www.enerdata.net/publications/daily-energy-news/china-central-asia-natural-gas-pipeline-supplied-43-bcm-china-2022.html#:~:text=In%202021%2C%20China%20imported%20a,%2C%20and%2011%25%20from%20Kazakhstan.
17 F. Umbach, "The South China Sea Disputes: The Energy Dimensions," *ISPSW Strategy Series: Focus on Defense and International Security* 494 (2017).
18 Green Finance & Development Center. *Countries of the Belt and Road Initiative (BRI)*. 2022. https://greenfdc.org/countries-of-the-belt-and-road-initiative-bri/
19 Report on the Findings and Recommendations from the Meetings of the Advisory Council of the Belt and Road Forum for International Cooperation in 2019 and 2020. Retrieved from http://ua.china-embassy.gov.cn/ydyls/202112/P020211222325476959329.pdf
20 "China Defers Tonga's Loan Payments as Pacific Nation Signs Up to Belt and Road." *Reuters*, November 18, 2018. Retrieved from https://www.reuters.com/article/us-apec-summit-tonga/china-defers-tongas-loan-payments-as-pacific-nation-signs-up-to-belt-and-road-idUSKCN1NN01J
21 "China's Xi Says the Belt and Road Must Be Green, Sustainable." *Reuters*, April 27, 2019. Retrieved from https://www.reuters.com/article/us-china-silkroad/chinas-xi-says-belt-and-road-must-be-green-sustainable-idUSKCN1S104I
22 OECD. *China's Belt and Road Initiative in the Global Trade, Investment and Finance Landscape*. (2018). Retrieved from https://www.oecd.org/finance/Chinas-Belt-and-Road-Initiative-in-the-global-trade-investment-and-finance-landscape.pdf
23 Z. Vörös, "PR China's Pursuit of a Blue-Water Navy," in *China in the Light of Global Challenges* (Budapest, CT: ELTE Konfuciusz Intézet (ELTE Confucius Institute), 2017).
24 Y. Zhao, X. Liu, S. Wang, and Y. Ge, "Energy Relations between China and the Countries along the Belt and Road: An Analysis of the Distribution of Energy Resources and Interdependence Relationships," *Renewable and Sustainable Energy Reviews* 107 (2019): pp. 133–144. doi:10.1016/j.rser.2019.03.007
25 "China's Great Game: In Russia's Backyard." *Financial Times*, October 14, 2015. Retrieved from https://www.ft.com/content/d35d34ca-6e70-11e5-aca9-d87542bf8673; Amy Jaffe and Robert Manning, "Russia, Energy and the West," *Survival* 43, no. 2 (2001): pp. 133–152. doi:10.1080/713660353
26 M. Laurelle (ed.), *China's Belt and Road Initiative and Its Impact in Central Asia* (Washington, DC: GW's Central Asia Program and Nazarbayev University, 2018), p. x.

27 These six economic corridors are the China-Mongolia-Russia Economic Corridor, the New Eurasian Land Bridge, the China-Central Asia-Western Asia Corridor, the China-Indochina Peninsula, the China-Pakistan Economic Corridor, and the Bangladesh-China-India-Myanmar Corridor.
28 "Why Central Asia Is Betting on China's Belt and Road." *The Diplomat*, August 13, 2018. Retrieved from http://www.eurasiancouncilforeignaffairs.eu/wp-content/uploads/2018/08/Why-Central-Asia-Is-Betting-on-China.pdf
29 X. Chen and F. Fazilov, "China and Central Asia: A Significant New Energy Nexus," *European Financial Review*.
30 M. Öğütçü, *China's Belt and Road Initiative: How Has China's Energy Security Changed over the Past Decade?* 19/33 (2019). Retrieved from Policy Center for the New South website: https://www.policycenter.ma/sites/default/files/PB%20-%2033-19%20%28Mehmet%29.pdf
31 Philippe Le Corre, *Kazakhstan Must Look Beyond the Belt and Road*, May 4, 2018. Retrieved from https://carnegieendowment.org/2018/05/04/kazakhstan-must-look-beyond-belt-and-road-pub-76266
32 Jinhui Duan, Shuying Wei, Ming Zeng, and Yanfang Ju, *The Energy Industry in Xinjiang, China: Potential, Problems, and Solutions*. Xinhua Net., October 3, 2019. Retrieved from https://www.powermag.com/energy-industry-xinjiang-china-potential-problems-solutions-web/
33 "Large Gas Field Discovered in Xinjiang." https://www.xinhuanet.com/english/2019-10/03/c_138446702.htm
34 The 13th Five-Year Plan for Economic and Social Development of the People's Republic of China (2016–2020). Retrieved from https://en.ndrc.gov.cn/policies/202105/P020210527785800103339.pdf
35 C. P. Freeman, "New Strategies for an Old Rivalry? China-Russia Relations in Central Asia after the Energy Boom," *The Pacific Review* 31, no. 5 (2018): pp. 635–654. doi:10.1080/09512748.2017.1398775
36 The table is based on International Trade Center Data (2023) for 2020 and 2021. *Trade Map: Trade Statistics for International Business Development Monthly, Quarterly and Yearly Trade Data*. The values are indicated in US Dollar Thousand. /intracen.org Retrieved from https://intracen.org/
37 "China Provides $585 million to Support Oil and Gas Projects in Uzbekistan." *Asia Plus*, October 23, 2019. Retrieved from https://www.asiaplustj.info/en/news/centralasia/20191023/china-provides-585-million-to-support-oil-and-gas-projects-in-uzbekistan
38 International Energy Agency. *Turkmenistan*. https://www.iea.org/countries/turkmenistan
39 CACI Analyst, "China's Military Base in Tajikistan: What Does It Mean?" April 18, 2019. Retrieved January 14, 2020, from https://www.cacianalyst.org/publications/analytical-articles/item/13569-chinas-military-base-in-tajikistan-what-does-it-mean?.html
40 The Observatory of Economic Complexity (OEC), OEC Country Profiles: Kazakhstan, Kyrgyzstan, Tajikistan, Turkmenistan and Uzbekistan. https://oec.world/
41 "Why Is Central Asia Betting on China's Belt and Road?" *The Diplomat*, August 13, 2018). Retrieved from http://www.eurasiancouncilforeignaffairs.eu/wp-content/uploads/2018/08/Why-Central-Asia-Is-Betting-on-China.pdf
42 N. Kassenova, "China as an Emerging Donor in Tajikistan and Kyrgyzstan." *Russie.Nei.Visions*, Issue 36, Paris: Ifri Russia/NIS Center, 2009.
43 "China Signs a Deal for 30 Years of Turkmen Gas." *The Guardian*, April 12, 2017. https://www.theguardian.com/business/2009/jun/25/china-turkmenistan-gas
44 N. Kassenova, *China as an Emerging Donor in Tajikistan and Kyrgyzstan*, 2009.
45 S. Custer, T. Sethi, J. Solis, J. Lin, S. Ghose, A. Gupta, R. Knight, and A. Baehr. *Silk Road Diplomacy: Deconstructing Beijing's Toolkit to Influence South and Central Asia*, December 10, 2019. (Williamsburg, VA: AidData at William & Mary, 2019).

46 "China's Surveillance State Has Eyes on Central Asia." *Foreign Policy*, November 15, 2019. https://foreignpolicy.com/2019/11/15/huawei-xinjiang-kazakhstan-uzbekistan-china-surveillance-state-eyes-central-asia/
47 "Partnership Is Much Better for China Than It Is for Russia." *The Economist*, July 27, 2019. Retrieved from https://www.economist.com/printedition/2019-07-27
48 M. Sharma, "India's Approach to China's Belt and Road Initiative—Opportunities and Concerns," *The Chinese Journal of Global Governance*, 2019. https://brill.com/view/journals/cjgg/5/2/article-p136_3.xml?language=en#note-FN000055
49 For a comprehensive account see, Edward Ashbee, *Countering China*, 2023.
50 "China's Belt and Road Initiative and Its Impact in Central Asia," *Voices on Central Asia*, 2018. Retrieved from https://voicesoncentralasia.org/chinas-belt-and-road-initiative-and-its-impact-in-central-asia/
51 John Hurley, Scott Morris, and Gailyn Portelance, *Examining the Debt Implications of the Belt and Road Initiative from a Policy Perspective*, 2018. Retrieved from Center for Global Development website: https://www.cgdev.org/sites/default/files/examining-debt-implications-belt-and-road-initiative-policy-perspective.pdf
52 Melinda Davies and Nathanial Matthews, "Water Futures along China's Belt and Road Initiative in Central Asia," *International Journal of Water Resources Development* 37, no. 6 (2021): pp. 955–975.
53 Joanna Lillis, "Uzbekistan Leader Warns of Water Wars in Central Asia," *Eurasia Net*, September 7, 2012. https://eurasianet.org/uzbekistan-leader-warns-of-water-wars-in-central-asia.
54 Melinda Davies and Nathanial Matthews, "Water Futures along China's Belt and Road Initiative in Central Asia," (2021): p. 960.
55 Elena F. Tracy, Evgeny Shvarts, Eugene Simonov, and Mikhail Babenko, "China's New Eurasian Ambitions: The Environmental Risks of the Silk Road Economic Belt," *Eurasian Geography and Economics* 58, no. 1 (2017): pp. 56–88. doi:10.1080/15387216.2017.1295876
56 Ibid.
57 Payrav Chorshanbiev, "Таджикистан постепенно превращается в цементный завод для соседних стран Подробнее: "[Tajikistan Is Gradually Turning into a Cement Plant for Neighboring Countries]. https://www.asiaplustj.info/ru/news/tajikistan/economic/20201026/donald-tramp-dosrochno-progolosoval-na-viborah-prezidenta-ssha-za-parnya-po-imeni-tramp, *Asia-Plus*, October 26, 2020.
58 I. Hofman, "Politics or Profits along the 'Silk Road': What Drives Chinese Farms in Tajikistan and Helps Them Thrive?" *Eurasian Geography and Economics* 57, no. 3 (2016): pp. 457–481. https://doi.org/10.1080/ 15387216.2016.1238313
59 "What Does China's BRI Mean for Central Asia's Water Future." *Eurasianet*, February 11, 2021. https://eurasianet.org/what-does-chinas-bri-mean-for-central-asias-water-future.
60 United Nations Economic and Social Commission for Asia and the Pacific (UNESCAP), *The Belt and Road Initiative for Seamless Connectivity and Sustainable Development in the Asia-Pacific Region*, 2021. https://repository.unescap.org/handle/20.500.12870/4099

References

Ashbee, Edward. *Countering China: US Responses to the Belt and Road Initiative*. Boulder, CO: Lynne Rienner Publishers, 2023. https://doi.org/10.1515/9781685859442

Asia-Plus. "China Provides $585 million to Support Oil and Gas Projects in Uzbekistan." October 23, 2019. https://www.asiaplustj.info/en/news/centralasia/20191023/china-provides-585-million-to-support-oil-and-gas-projects-in-uzbekistan

Bousso, Ron. "As Renewables Soar, BP Sees China Hitting Brakes on Energy Growth." *Reuters*. February 14, 2019. Retrieved from https://www.reuters.com/article/us-bp-energy/as-renewables-soar-bp-sees-china-hitting-brakes-on-energy-growth-idUSKCN1Q3244.

BP. "China's Energy Market in 2018." *BP Statistical Review*. 2019. https://www.bp.com/content/dam/bp/business-sites/en/global/corporate/pdfs/energy-economics/statistical-review/bp-stats-review-2019-china-insights.pdf

CACI Analyst. "China's Military Base in Tajikistan: What Does It Mean?" April 18, 2019. Retrieved January 14, 2020, from https://www.cacianalyst.org/publications/analytical-articles/item/13569-chinas-military-base-in-tajikistan-what-does-it-mean?.html.

Cadell, C., and B. Goh. "China's Xi Says the Belt and Road Must Be Green, Sustainable." *Reuters*. April 27, 2019. Retrieved from https://www.reuters.com/article/us-china-silkroad/chinas-xi-says-belt-and-road-must-be-green-sustainable-idUSKCN1S104I.

CECEP. "Xi Jinping: Promote China's Energy Production and Consumption Revolution Vigorously." June 27, 2014. Accessed April 29, 2023. http://www.cecep.cn/g3621/s12533/t37788.aspx.

Chen, X., and F. Fazilov. "China and Central Asia: A Significant New Energy Nexus." *European Financial Review*. 2013.

Chorshanbiev, Payrav. "Таджикистан постепенно превращается в цементный завод для соседних стран Подробнее: " [Tajikistan Is Gradually Turning into a Cement Plant for Neighboring Countries]. *Asia-Plus*. 26 October, 2020. https://www.asiaplustj.info/ru/news/tajikistan/economic/20201026/donald-tramp-dosrochno-progolosoval-na-viborah-prezidenta-ssha-za-parnya-po-imeni-tramp.

Cooley, A. *Great Games, Local Rules: The New Power Contest in Central Asia*. New York: Oxford University Press, 2012.

Cooley, Alexander, and Daniel H. Nexon. *Exit from Hegemony: The Unraveling of the American Global Order*. New York: Oxford University Press, 2020.

Custer, S., T. Sethi, J. Solis, J., Lin, S. Ghose, A. Gupta, R. Knight, and A. Baehr. "Silk Road Diplomacy: Deconstructing Beijing's Toolkit to Influence South and Central Asia." Williamsburg: AidData at William & Mary, 2019.

Daojiong, Z. "China's Energy Security: Domestic and International Issues." *Survival* 48, no. 1 (2006): 179–190. doi:10.1080/00396330600594322

Davies, Melinda, and Nathanial Matthews. "Water Futures along China's Belt and Road Initiative in Central Asia." *International Journal of Water Resources Development* 37, no. 6 (2021): 955–975. doi: 10.1080/07900627.2020.1856049

Duan, Jinhui, Shuying Wei, Ming Zeng, and Yanfang Ju. "The Energy Industry in Xinjiang, China: Potential, Problems, and Solutions." January 1, 2016. Retrieved from https://www.powermag.com/energy-industry-xinjiang-china-potential-problems-solutions-web/.

Enerdata. "China-Central Asia Natural Gas Pipeline Supplied 43 bcm to China." https://www.enerdata.net/publications/daily-energy-news/china-central-asia-natural-gas-pipeline-supplied-43-bcm-china-2022.html#:~:text=In%202021%2C%20China%20imported%20a,%2C%20and%2011%25%20from%20Kazakhstan.

Eurasianet. "What Does China's BRI Mean for Central Asia's Water Future." February 11, 2021. https://eurasianet.org/what-does-chinas-bri-mean-for-central-asias-water-future.

Financial Times. "China Now World's Biggest Energy User." July 19, 2010. https://www.ft.com/content/937fdd2c-934b-11df-bb9a-00144feab49a.

Financial Times. "China's Great Game: In Russia's Backyard." October 14, 2015. https://www.ft.com/content/d35d34ca-6e70-11e5-aca9-d87542bf8673

Freeman, C. P. "New Strategies for an Old Rivalry? China-Russia Relations in Central Asia after the Energy Boom." *The Pacific Review* 31, no. 5 (2018): 635–654. doi:10.1080/09512748.2017.1398775

General Administration of Customs China (GACC). *China Customs Statistics*: Coverage of Major Imports and Exports. http://english.customs.gov.cn/statistics/Statistics?ColumnId=6.

Greenfield, Charlotte. "China Defers Tonga's Loan Payments as Pacific Nation Signs Up to Belt and Road." *Reuters*. November 18, 2018. Retrieved from https://www.reuters.com/article/us-apec-summit-tonga/china-defers-tongas-loan-payments-as-pacific-nation-signs-up-to-belt-and-road-idUSKCN1NN01J.

Harding, Luke. "China Signs a Deal for 30 Years of Turkmen Gas." *The Guardian*. June 25, 2009. https://www.theguardian.com/business/2009/jun/25/china-turkmenistan-gas.

Hofman, I. "Politics or Profits along the 'Silk Road': What Drives Chinese Farms in Tajikistan and Helps Them Thrive?" *Eurasian Geography and Economics* 57, no. 3 (2016): 457–481. https://doi.org/10.1080/ 15387216.2016.1238313.

Hurley J., Scott Morris, and Gailyn Portelance. "Examining the Debt Implications of the Belt and Road Initiative from a Policy Perspective." Retrieved from Center for Global Development, 2018. https://www.cgdev.org/sites/default/files/examining-debt-implications-belt-and-road-initiative-policy-perspective.pdf.

Jardine, Bradley. "China's Surveillance State Has Eyes on Central Asia." *Foreign Policy*. November 15, 2019. https://foreignpolicy.com/2019/11/15/huawei-xinjiang-kazakhstan-uzbekistan-china-surveillance-state-eyes-central-asia/.

Kassenova, N. "China as an Emerging Donor in Tajikistan and Kyrgyzstan." Russie.Nei. Visions, Issue 36, Paris: Ifri Russia/NIS Center, 2009. https://www.ifri.org/sites/default/files/atoms/files/ifrichinacentralasiakassenovaengjanuary2008.pdf.

Laurelle, M. "China's Belt and Road Initiative and Its Impact in Central Asia." *Voices on Central Asia*. January 19, 2018. Retrieved from https://voicesoncentralasia.org/chinas-belt-and-road-initiative-and-its-impact-in-central-asia/.

Laurelle, M. (ed.), "China's Belt and Road Initiative and Its Impact in Central Asia." Washington DC: GW's Central Asia Program and Nazarbayev University, 2018. https://centralasiaprogram.org/wp-content/uploads/2017/12/OBOR_Book_.pdf.

Le Corre, Philippe. "Kazakhstan Must Look Beyond the Belt and Road." Carnegie Endowment. May 4, 2018. https://carnegieendowment.org/2018/05/04/kazakhstan-must-look-beyond-belt-and-road-pub-76266

Lillis, Joanna. "Uzbekistan Leader Warns of Water Wars in Central Asia." Eurasia Net. September 7, 2012. https://eurasianet.org/uzbekistan-leader-warns-of-water-wars-in-central-asia.

Nedopil, Christoph. "Countries of the Belt and Road Initiative (BRI)." Shanghai, Green Finance & Development Center, FISF Fudan University, 2023. https://greenfdc.org/countries-of-the-belt-and-road-initiative-bri/.

OECD. "China's Belt and Road Initiative in the Global Trade, Investment and Finance Landscape." 2018. https://www.oecd.org/finance/Chinas-Belt-and-Road-Initiative-in-the-global-trade-investment-and-finance-landscape.pdf

Öğütçü, M. "China's Belt and Road Initiative: How Has China's Energy Security Changed Over the Past Decade?" Retrieved from Policy Center for the New South, 2019. https://www.policycenter.ma/sites/default/files/PB%20-%2033-19%20%28Mehmet%29.pdf.

Report on the Findings and Recommendations from the Meetings of the Advisory Council of the Belt and Road Forum for International Cooperation in 2019 and 2020. Retrieved from http://ua.china-embassy.gov.cn/ydyls/202112/P020211222325476959329.pdf

Sharma, M. "India's Approach to China's Belt and Road Initiative—Opportunities and Concerns." *The Chinese Journal of Global Governance* (2019). https://brill.com/view/journals/cjgg/5/2/article-p136_3.xml?language=en#note-FN000055.

The Diplomat. "Why Central Asia Is Betting on China's Belt and Road." August 13, 2018. Retrieved from http://www.eurasiancouncilforeignaffairs.eu/wp-content/uploads/2018/08/Why-Central-Asia-Is-Betting-on-China.pdf

The Economist. "Partnership Is Much Better for China Than It Is for Russia." July 27, 2019. Retrieved from https://www.economist.com/printedition/2019-07-27.

The Observatory of Economic Complexity (OEC), "OEC Country Profiles: China, Crude Petroleum." 2023. https://oec.world/en/profile/bilateral-product/crude-petroleum/reporter/chn

The Observatory of Economic Complexity (OEC), "OEC Country Profiles: Kazakhstan, Kyrgyzstan, Tajikistan, Turkmenistan and Uzbekistan." 2023. https://oec.world.

Tracy, E. F., Evgeny Shvarts, Eugene Simonov, and Mikhail Babenko. "China's New Eurasian Ambitions: The Environmental Risks of the Silk Road Economic Belt." *Eurasian Geography and Economics* 58, no. 1 (2017): 56–88. doi: 10.1080/15387216.2017.1295876.

Trade Map: Trade Statistics for International Business Development Monthly, Quarterly and Yearly Trade Data. Import & Export Values, Volumes, Growth Rates, Market Shares, etc. Retrieved from https://intracen.org/.

Umbach, F. "The South China Sea Disputes: The Energy Dimensions." *ISPSW Strategy Series: Focus on Defense and International Security*, 494. July 2017. https://ethz.ch/content/dam/ethz/special-interest/gess/cis/center-for-securities-studies/resources/docs/ISPSW-494_Umbach%20(2).pdf.

United Nations Economic and Social Commission for Asia and the Pacific (UNESCAP). "The Belt and Road Initiative for Seamless Connectivity and Sustainable Development in the Asia-Pacific Region." 2021. https://repository.unescap.org/handle/20.500.12870/4099.

Vörös, Z. PR China's Pursuit of a Blue-Water Navy. In China in the Light of Global Challenges. Budapest, CT 2017: ELTE Konfuciusz Intézet (ELTE Confucius Institute).

World Bank. (n.d.). *The World Bank in China: Overview*. Retrieved from https://www.worldbank.org/en/country/china/overview.

Xinhua Net. "Large Gas Field Discovered in Xinjiang." October 3, 2019. https://www.xinhuanet.com/english/2019-10/03/c_138446702.htm.

Yongqi, Hu. "Three-Year Action Plan Aims for Blue Skies." *China Daily*. June 19, 2018. Retrieved from https://www.chinadaily.com.cn/a/201806/19/WS5b287383a310010f8f59d939.html

Zhao, H. *The Economics and Politics of China's Energy Security Transition*. Cambridge, MA: Academic Press, 2018.

Zhao, Y., X. Liu, S. Wang, and Y. Ge. "Energy Relations Between China and the Countries Along the Belt and Road: An Analysis of the Distribution of Energy Resources and Interdependence Relationships." *Renewable and Sustainable Energy Reviews* 107, (2019): 133–144. doi:10.1016/j.rser.2019.03.007.

7

TÜRKİYE AND THE ENERGY AND WATER SUPPLY NETWORK IN THE SOUTH CAUCASUS AND THE GREATER MIDDLE EAST

In the post-Cold War era, Türkiye has emerged as a critical regional actor, whose policies have significant implications for the Caucasus and Eurasia region. In addition to Türkiye's energy ties and geopolitical interests, its foreign policy towards this region and the greater Middle East are shaped by identity based and ideational factors as well. While Türkiye's more assertive policy stance towards this region reflects the pan-Turkist and neo-Ottomanist trends in Turkish foreign policy as argued by some scholars,[1] we posit that the economic ties and energy interests emerge as prevalent factors shaping the relations. Hence, Ankara has close links with this region in relation to the energy-water nexus. Moreover, the energy and water related policies also have significant implications for food security.

The South Caucasus region and Türkiye experience persistent challenges in relation to the Eurasian energy-water nexus. This chapter examines the domestic and international dynamics of their energy and water security issues and policies. In this context, in line with our theoretical framework energy, environment and security related factors within a volatile geopolitical context will be analyzed. Energy and water politics constitute indispensable parts of natural resource security, which in turn are strongly linked to human security. This situation has significant impact for the South Caucasus particularly regarding their ties with Türkiye and Russia—and with each other. We argue that these countries suffer from the legacy of Soviet environmental neglect and disregard, and while there are important degrees of remedial progress, there remain severe challenges and missed opportunities. These factors can be explained in large part in terms of more general challenges at the domestic level in the political and economic realm, including the lack of transparency, inclusivity, and effective governance of transboundary resources, coupled with the apparent tendency to favor other development priorities over sustainability.

DOI: 10.4324/9781003373735-7

Moreover, at the international level, geopolitical fault lines and regional conflicts such as the Russian invasion of Ukraine exacerbate these challenges.

Domestic-External Linkages: The Soviet Legacy and Russian Impact in the South Caucasus

Azerbaijan regained political independence in 1991 with the collapse of the Soviet Union. The early years were troubled and uncertain, as the relationship with neighboring Armenia grew poisonous from the war over Nagorno-Karabakh. Russian policy towards this region alternated between meddling and neglect. Heydar Aliyev (former First Secretary of the Communist Party in Azerbaijan from 1969 to 1982) came to power in 1993. He was able to consolidate his rule largely by 1995 and subsequently proved effective in walking the tight rope between lingering Russian influence and the interests of Western governments and corporations eager to develop the oil and gas fields with independent transportation links. The development of a special relationship with Türkiye with strong US backing was critical in this balance. Economic independence and relative prosperity went hand in hand; the question remains as to whether the energy/resource curse will prevail or whether the economy will become diversified enough to remain viable after the oil and gas reserves run out. As a more immediate challenge, the Covid-19 pandemic and regional conflicts have also intensified concerns regarding energy price volatility.

The oil and gas industry are managed by SOCAR, the State Oil Company of the Azerbaijan Republic, which has close ties to the President. SOCAR has inherited infrastructure and environmental impact from the Soviet energy development and construction apparatus—most notably apparent in the offshore drilling installation and support city: Neft Daslari—Oily Rocks—built in 1949 some 60 miles out from Baku in the Caspian Sea. Aging equipment, inefficient production, and scant concern about environmental impact are the Soviet legacy—one hard to fix despite SOCAR's ability to attract foreign investment for energy development and the substantial income thus far earned from oil and gas exports.[2]

Post-independence energy projects include the Azeri-Chirag-Guneshli oil field complex (discovered in 1985, but with production beginning only in 1997) and the Shah Deniz gas field (discovered in 1999). Both fields are operated by a consortium of companies and agencies led by British Petroleum and are to the East and Southeast of Baku in the Caspian Sea. In recent years, Azerbaijan's oil exports have exceeded 718,000 barrels per day. Its natural gas exports have exceeded 8 billion cubic meters per year. Azerbaijan ranks 19th and 24th, respectively, among world exporters in oil and natural gas.[3]

In 2003, President Heydar Aliyev was succeeded by his son Ilham Aliyev, when the former President left office ostensibly due to failing health. Ilham Aliyev has demonstrated a skillful balancing act between augmenting Azerbaijan's relations with the West without alienating Russia. Azerbaijan was a founding member of the CIS, but it was also a founding member of Organization for Democracy and Economic Development (GUAM) and is an active member of NATO's Partnership for Peace. It has contributed troops to Western-led military and peacekeeping operations in Kosovo, Iraq, and Afghanistan. Along with Georgia, Azerbaijan joined the Collective Security Treaty Organization (CSTO) in 1994, but both withdrew in 1999. It was also that year that a prominent advisor to the President, Vafa Gulazade, urged that NATO establish a military base in Azerbaijan.[4] This was not taken seriously by US officials who at that point were anxious not to prematurely disrupt the prospects for a strategic partnership with Russia.

Moscow's military backing for Armenia during the first Nagorno-Karabakh War and its subsequent reluctance to pressure Yerevan for reaching an internationally recognized settlement increased concerns over Russian motives. Russia remains one of the few states that does not acknowledge the Azerbaijani claim to the territory occupied by Armenia, but it also did not veto UN Security Council Resolutions 822, 853, 874, and 884 which called for immediate withdrawal of Armenian forces from all occupied Azerbaijani land.

Heydar Aliyev managed to navigate beyond Russia's assertive stance of veto rights over energy deals in the Caspian region, when on September 20, 1994, he signed the "Contract of the Century" for development of offshore wells near Baku.[5] By September 2000, the Baku-Tbilisi-Ceyhan (BTC) pipeline was approved to ensure oil shipments, independent from Russian territorial control to the Southeastern Mediterranean coast of Türkiye. The first oil left Baku on May 10, 2005 reaching Ceyhan. This effectively ended the Russian oil transportation monopoly for Caspian shipments and opened a new era of Russia's relationship with the region. This did not, however, lead to regional harmony or environmental cooperation. Azerbaijan has not joined the Eurasian customs union pushed by President Putin, but it has also not signed an Association Agreement with the EU. Azerbaijani leaders signed a Partnership and Cooperation Agreement (PCA) with the EU in 1996. The government adopted the European Neighborhood Policy Action plan on November 14, 2006. The relationship includes considerable development assistance and cooperation on energy through the EU's Interstate Oil and Gas Transportation to Europe program (INOGATE) for enhancing energy security, convergence of member state energy markets, supporting sustainable energy development, and attracting investment for energy projects.

Georgia had a tumultuous decade after gaining its independence in 1991. During these early years, Russia played an active role, as much as its own turmoil at home permitted. The political legitimacy of the Presidential regime of former Soviet Foreign Minister Eduard Shevardnadze swiftly eroded after his re-election in 2000. In this period, the then Russian Foreign Minister Igor Ivanov assumed a significant mediating role between Shevardnadze and the opposition during the "Rose Revolution," which ultimately brought Mikhail Saakashvili into political power.

Georgia has been an eager partner of the BTC pipeline project. Lacking fossil fuel reserves of its own and being persistently anxious about Russian engagement in the region, the Georgian government viewed this project not only as a significant economic connection to Azerbaijan, Türkiye, and the EU but also as a foreign policy move consolidating its ties with the West. In the meantime, Georgia's domestic energy policy concentrated on expanding and modernizing its hydroelectric power generation. Unfortunately, this was mostly done without due diligence to the environmental effects, often following the Soviet legacy of environmental neglect. Georgia currently exports electricity to Türkiye and even Russia, at least in the periods when the rivers flow swiftly. This began with revitalization of the Soviet era plant at the Enguri dam, with more than $50 million of funding from the World Bank, EU, EBRD, and the Japanese government. The Chinese firm, Sino Hydro, provided some of the technical and construction assistance. Future plans include 15 new hydroelectric plants in the near term. The search for additional ways to develop the economy and to restore agricultural productivity, however, remains a daunting task.

The 2008 war with Russia over South Ossetia and Abkhazia turned out to be a dismal strategic move. It created an opportunity for Putin to consolidate his authority in the region, particularly at President Saakashvili's expense. However, it was not a convincing display of power, and it also significantly undermined Russia's international standing.

In October 2012, Bidzina Ivanishvili came to power as Prime Minister leading the "Georgian Dream" coalition of former opposition parties. Ivanishvili was viewed to be more amiable towards Russia than President Mikheil Saakashvili. Nevertheless, the new Prime Minister had promised during his election campaign that "My first visit will be to Washington and the United States is our main partner." He also stated that his goal was to enhance Georgia's ties with Russia while also seeking membership in NATO.[6] The troubled Georgia-Russia relations had in fact somewhat improved before the elections, as the Georgian government strove for better trade relations and higher Russian tourist numbers. Tourism received a boost with the institution of a visa-free regime for Russian visitors in July of 2011. The 2014 crisis in Ukraine, however, once again undermined the geopolitical dimension of the relationship.

Ivanishvili's Georgian Dream coalition also prevailed in the Presidential elections of October 2013. Giorgi Margvelashvili then replaced Saakashvili who was term limited, and likely not that interested in a Presidency that would now have less power vis-à-vis the Prime Minister, given structural reforms in place at his direction. The Russian press welcomed his departure, sometimes echoing comments from the Kremlin that questioned his sanity. Bidzina Ivanishvili then followed through on his promise to step down. Irakli Garibashvili became Prime Minister on November 20, 2013. As a former business leader associated with Ivanishvili, he had previously served as Minister of Internal Affairs in his administration.

A major new thrust in Georgian development policies resulting from this leadership change was an effort to begin rebuilding the collapsed agricultural sector.

Despite its prominence during the Soviet period, it had been neglected essentially since the chaos of the early years of independence. Saakashvili saw the future of Georgia as one of prosperity through economic liberalization and new enterprise development in the cities. Giorgi Kadagidze, head of the Georgian National Bank between 2009 and 2016, was reported widely to say that the solution to rural poverty was for the rural population to move to cities, especially the proposed new Black Sea port city. But the poverty of the rural population grew, and an agricultural sector that had supplied much of the fruits, vegetables, and wine consumed by the Soviet Union was now unable to meet even the needs of the Georgian population with dismal implications at the local level for food security. Food imports from Türkiye, Azerbaijan, and even Armenia became the norm.

Rebuilding the agricultural sector will take time and investment. The Agricultural University of Georgia deteriorated from its former prominence—privatized in 2011 under somewhat controversial circumstances and transformed to focus more on engineering and management. Extension of services to rural populations for technical assistance must be built from scratch; the agencies/offices exist, but they are largely not staffed at a time when basic farming know-how seems to have been lost in many regions. To address these challenges, according to the *Strategy for Agricultural Development in Georgia, 2015–2020*, "New municipal information and consultancy centers have been created to improve communication with farmers, define priorities and provide support services. To improve the working process and services, the needs of farmers and stakeholders will be considered to the fullest possible extent."[7] Most serious challenge, however, has been the deterioration in agricultural land in key regions. With the dissolution of collective farms after independence, small plots are the most common feature, and while there are a variety of producer associations, and a recent law on agricultural cooperation, collaboration between small holders to create economies of scale and effective mechanization remains limited.

Bidzina Ivanishvili, who was Prime Minister of Georgia from 2012 to 2013, established an agricultural development fund shortly after his election, reportedly with a large contribution from his own personal wealth. Agricultural development was also a priority of the US Millennium Challenge Corporation's second compact with the Republic of Georgia, signed in June of 2013.

Georgia also faces serious water quality challenges. There is abundant fresh water from its many rivers, but it is not distributed evenly beyond the mountainous West, and there are serious pollution patterns from mining operations and industrial dumping. The larger cities, especially Tbilisi, Kutaisi, and Rustavi, have struggled with outdated or dysfunctional water and sewage treatment systems. The environmental crisis threatening the Black Sea also plagues Georgia, particularly its coastal cities and wetlands (e.g., Ramsar), after decades of unrestrained sewage flows, industrial pollution, and agricultural chemical runoff from the littoral countries and rivers sourced in Eastern Europe and Russia, oil spills, and waste dumping by the large-scale tanker traffic.[8]

There have also been promising initiatives such as the Kura *II Project*, which targets the enhancement of cooperation between Azerbaijan and Georgia in water resources management of the Kura River. The Global Environment Facility (GEF) and United Nations Development Program (UNDP) supported this four-year project with an allocated budget of 5.3 million USD. In return, national partners in Azerbaijan and Georgia have contributed over $190 million in co-financing. The goal is to deepen cooperation between water-using sectors from Azerbaijan and Georgia to work towards integration and coordination of water use plans within and between the countries. Integrated Water Resources Management (IWRM) and implementation of more effective local governance mechanisms would be critical for these countries to be able to survive and thrive under climate change.[9]

In the South Caucasus, Armenia confronts the greatest challenge to find a viable strategy to meet its resource needs on its own, given that it has neither the copious river systems of Georgia nor the oil and gas reserves of Azerbaijan. What it does have is substantial support from the Armenian diaspora communities in Europe and the United States and foreign military and political assistance from Russia as needed. Russia's policy toward Armenia under Putin seems to reflect a continuing interest in preserving the status quo. Despite being entrusted with an important responsibility for resolving the conflict with Azerbaijan over Nagorno-Karabakh, one can point to little concrete effort to promote a negotiated settlement. The domestic imperatives in Armenia meanwhile seem compatible with this Russian posture and intent on continued reliance on an outdated (Soviet designed) nuclear power plant located on a geological fault line, and a policy of exploitation of water resources in contested areas causing further tensions with Azerbaijan.

Ironically, as with much of the larger region, those with adequate energy resources are without adequate water and vice versa. For instance, Azerbaijan, as a net energy-exporter country, has scarce water resources, while Georgia has abundant water yet lacks energy resources. Armenia has challenges regarding both, which increases its dependence on Russia for energy and transboundary river systems for water.

Russia's relations with Armenia took a turn in 2013 in connection with the effort to expand the customs union established initially between Russia, Belarus, and Kazakhstan. Faced with the choice between signing an Association Agreement with the European Union and joining the Customs Union, Armenia ultimately chose the latter. President Putin then visited Armenia on December 2 for implementation discussions. Twelve agreements between the two governments were signed for economic, energy, and security cooperation. This seemed to ease Armenian concerns about reported Russian arms sales in 2011–2012 to Azerbaijan, and Armenia's search for a more independent course. But Armenian motives for "caving" to the Russian pressure reflected multiple concerns. Threats to agricultural exports to Russia may have been important, as well as the continuing relevance of political and military ties to Russia's only Caucasus ally, including the Russian 102nd base at Gyumri, and assurance of Armenian territorial integrity, ostensibly versus

Türkiye and Iran. EU concerns about the continued stationing of Russian troops in Armenia were expressed from time to time, and President Putin did declare that Armenia would have to choose.[10]

The fragile nature of the region regarding frozen conflicts risks energy supply security as witnessed during the Second Karabakh War in 2020. The armed conflict has spread to Azerbaijan's territories outside the disputed areas with strategic significance for natural gas and oil trade. The city of Tovuz, a major hub for natural gas and oil pipelines from Azerbaijan to Türkiye and all the way to Europe, was hit several times during the conflict which demonstrates the sensitivities of the East-West Energy Corridor. Despite the ceasefire, therefore, the continuous risk of re-escalation and even a full-scale war constitutes a major threat for energy routes passing through the Caucasus.

Environment and security priority areas in the South Caucasus are in proximity, and in terms of governance, there is a pressing need for regional collaboration for the management of cross-border environmental concerns. However, political and economic challenges on the domestic front, combined with a turbulent and conflict-laden international contexts, hinder much needed cooperation for sustainability and the establishment of essential governance structures for more effective utilization of transnational natural resources.

Türkiye's Relations with the South Caucasus: Implications for Energy and Water Security

Following the dissolution of the Soviet Union, pan-Turkist sentiments significantly increased in Türkiye. Since Azerbaijan and the Turkic Republics in Central Asia were part of the USSR, they were mostly beyond Türkiye's foreign policy reach during the Cold War era. However, consecutive declarations of independence created substantial excitement in Türkiye. As a key figure of the time as Prime Minister (1983–1991) and then as President (1991–1993), Turgut Özal as well as the nationalists in Türkiye regarded this as a tantalizing opportunity to advance a broader pan-Turkist agenda. Moreover, the goal was to promote Türkiye as a model in this region. The United States and the EU were also supportive to curb the competition from Russian and Iranian influence. Consequently, Ankara initiated an assertive campaign towards these newly founded Turkic states on the grounds of shared ethno-cultural identity.

Despite efforts such as the foundation of Turkish Cooperation and Coordination Agency (TİKA) and frequent high-level visits, the lack of adequate financial resources for supporting their state-building processes, as well as the resilience of the Russian influence in the region played a crucial role in limiting Türkiye's aims to expand its cooperation with these states into a pan-Turkist path. While emphasis on shared ethnic identity remains in the narratives, pan-Turkist elements relatively diminished regarding policies. Nevertheless, the cooperation between Türkiye and other Turkic states has been institutionalized through the Turkic Council and some other multilateral platforms. Moreover, as far as the South Caucasus is concerned strong ethnic and cultural ties with Azerbaijan, along with congruent energy and geopolitical interests serve as a key driver and consolidator of bilateral cooperation.

Ultimately, the pan-Turkist elements of the 1990s have been increasingly replaced with a neo-Ottomanist turn in Turkish foreign policy under the Justice and Development Party (AKP) rule, which came to power in 2002. Following an era of destructive wars in the Balkans resulting in the dissolution of Yugoslavia, Türkiye aimed at ensuring stability in the region with mediation efforts. Moreover, Türkiye actively engaged in the region through development coordination, preservation of historical and cultural heritages, and peacekeeping missions. Regarding the greater Middle East, on the other hand, the AKP government pursued a highly assertive policy as in the case of Palestinian conflict, as well as an over-engagement in the post-Arab Spring Middle East, and particularly in Syria. In the more recent period, as pressing economic and geopolitical problems started to dominate Türkiye's domestic and foreign policy agenda, Ankara has effectively moved into what Dalacoura called the "post-ideological" and more transactional phase of Turkish foreign policy regarding the greater Middle East.[11] In this context, while ethnic and religious ties maintains their significance, quest for energy security and geopolitics emerge as key pragmatic drivers shaping Türkiye's foreign policy towards these regions. Moreover, the climate change further accentuates the significance of hydropolitics in a drastically arid region with transboundary river systems.

In the South Caucasus, the most significant cooperation partner for Türkiye has been Azerbaijan. Bilateral relations have political, economic, and security dimensions in a complicated foreign policy context.[12] Moreover, as mentioned earlier, there has been a considerable emphasis on the common Turkish identity. In the volatile environment of Eurasia, Türkiye and Azerbaijan have been striving to increase their influence in the Eurasian energy game.

Given its rich energy resources, Azerbaijan has emerged as an important player in Eurasian energy politics. In 2019, Azerbaijani oil reserves constituted 0.4% of the world reserves and 1.4% of the global natural gas reserves.[13] In 2021, Azerbaijani share in global oil and natural gas production ranked 0.8%, respectively. Azerbaijan is 35th both in world oil and natural gas production.[14] While Azerbaijan's position in the energy sector is not as prevalent as some of the other Eurasian and the Middle Eastern producers such as Russia, Saudi Arabia, Qatar, and Iran; its prospects for future energy trade and diversification makes Azerbaijan an important player. The consumers of the Eurasian and the Middle Eastern energy, particularly Türkiye and European countries, are struggling to increase their energy security by diversifying their suppliers. For example, in natural gas, Türkiye's dependence on Russia and Iran was more than 69% in 2016,[15] while this number was 44.65% in 2020. The EU also has high dependence on Russian natural gas. In 2021, the EU imported 155 billion cubic meters of natural gas from Russia and consequently Russian energy imports accounted for around 45% of EU gas imports.[16] As discussed earlier, Russian invasion of Ukraine in March 2022 clearly revealed

EU vulnerabilities in terms energy security. In this context, Azerbaijan currently emerges as an important diversification source for Türkiye and the European countries especially in natural gas.

One of the first major energy initiatives in the region after the collapse of the Soviet Union and the independence of Azerbaijan was the Baku-Tbilisi-Ceyhan (BTC) pipeline project. With the strong involvement and backing of the US government, the project started in 2002 despite fierce competition from Russia and Iran. The construction of the pipeline was completed in 2005 and the first oil was pumped the next year. In this positive environment, Türkiye began purchasing natural gas from Azerbaijan in 2007. After this promising start, in the following years, energy relations between Türkiye and Azerbaijan have proceeded with ups and downs. Azerbaijan started to make more investments in the energy sector. In the meantime, Türkiye began its quest to become an energy hub, despite lacking substantial energy resources of its own. Türkiye has been trying to emerge as an energy bridge between rich oil and natural gas resources of the Caucasus and the European markets.[17] In this context, Türkiye's high levels of energy dependency constitute a major challenge.[18] For instance, Türkiye had 54 billion dollars of energy imports in 2011, which corresponded to 69% of Türkiye's rather problematic balance of accounts deficit.[19] In 2015, with lower oil prices Türkiye's energy import bill totaled $37.8 billion.[20] Figure 7.1 shows Türkiye's self-sufficiency by energy source between 2000 and 2019.

Turkish-Azerbaijani energy relations, especially after 2009, indicate that energy policy is closely linked to other foreign policy issues. On April 22, 2009, Türkiye and Armenia signed an agreement, which defined a provisional roadmap for the improvement of relations. This agreement created discomfort on the Azerbaijani side, which has problematic relations with Armenia especially due to the Nagorno-Karabakh conflict. Just three days after the signing of this agreement, the President of the state-owned oil and natural gas company, State Oil Company of the Republic of Azerbaijan (SOCAR), voiced Azerbaijani demands for

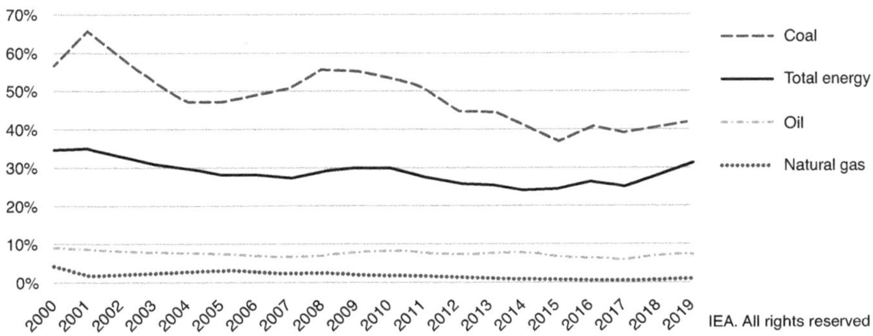

FIGURE 7.1 Self-Sufficiency by Energy Source, Türkiye (2000–2019).

Source: IEA (2020), IEA World Energy Statistics and Balances (database), www.iea.org/statistics.[21]

a new price deal in Turkish gas purchases from Azerbaijan.[22] The agreement that defined the old natural gas price had expired in 2008. Since then, Türkiye continued to purchase natural gas from Azerbaijan at the old prices. This price dispute created an unexpected crisis between the two countries. At the end, the Turkish Prime Minister Erdoğan had to visit Baku to alleviate the tensions.

Azerbaijani demands for a new price deal in the immediate aftermath of Turkish-Armenian rapprochement was diplomatic signaling reflecting the Azeri discontent by using the energy card. Moreover, Prime Minister Erdoğan's visit to Azerbaijan was not only to resolve the price dispute but also to mitigate political tensions. During this visit, Turkish side assured Baku that it would not violate Azerbaijani interests at the expense of improving its ties with Armenia. Ultimately, Türkiye and Azerbaijan settled their price dispute by an agreement on the Shah Deniz-II project in 2010 that defined new prices for the purchases, which fluctuates according to market conditions.[23]

After the price dispute, bilateral energy relations entered a new phase with the signing of the intergovernmental agreement for the Nabucco pipeline project on 13 July 2009. Nabucco pipeline aimed to take the Azeri natural gas to the Eastern and Central Europe through Türkiye as the fourth component of the East-West energy corridor after Baku-Tbilisi-Ceyhan (BTC) oil pipeline, the Baku-Tbilisi-Erzurum (BTE) natural gas pipeline, and the Türkiye-Greece gas interconnector line. Although the project was projected as a big opportunity for the European countries to diversify their energy suppliers and transit routes, there had been serious concerns about the project. Particularly, the lack of throughput commitment and the doubts about the amount of resource availability to finance and sustain the project caused hesitation for the parties. Ultimately, the EU partners failed to show determination and consensus to realize the project.[24] Moreover, Russia has been a major obstacle for the project as well. Since its monopoly on the East-West energy corridor was threatened, the Kremlin tried to undermine the Nabucco project by promoting the South Stream project as an alternative.[25]

Russian obstructions and uncertainties about the Nabucco pipeline project compelled Türkiye and Azerbaijan to seek new initiatives. Consequently, they signed a memorandum of understanding on the Trans-Anatolian pipeline project (TANAP) during the last week of 2011. According to this agreement, annually about 16 billion cubic meters natural gas will be transported. 6 billion cubic meters will be used in Türkiye and the remaining 10 billion cubic meters will be transferred to Europe. The capacity of gas flow is expected to reach to 31 billion cubic meters in 15 years. In June 2012, Turkish President Recep Tayyip Erdoğan and President of Azerbaijan Ilham Aliyev signed the deal to launch TANAP. Accordingly, the project was inaugurated on 12 June 2018.[26]

After the memorandum of understanding was signed, experts argued that the Trans-Anatolian pipeline would mean the end of the Nabucco project.[27] As anticipated, eventually the Nabucco pipeline project was shelved. Instead, Azerbaijan,

Türkiye, and their European partners agreed that Azeri gas will be transferred to Europe via TANAP, which is a less ambitious but a more feasible project than Nabucco.[28] In 2013, there was a final decision that the remaining part of the pipeline network will be connected to the Trans-Adriatic pipeline (TAP).[29] Hence, TANAP agreement created new opportunities for Azerbaijan, Türkiye, and European countries for energy cooperation. Figure 7.2 illustrates the current pipelines crossing the region indicating its critical geopolitical location as an energy crossroads.

These energy initiatives necessitated Ankara to appease Moscow as well. Russia is the leading natural gas supplier for Türkiye with more than 33% of Turkish gas imports coming from Russia.[31] Russia views these new projects as a potential obstacle for Russian domination in the Turkish and European energy markets. Hence, to ease the tensions, Ankara also signed an agreement with Moscow to transport Russian natural gas to Europe via South Stream under the Turkish Exclusive Economic Zone in the Black Sea. The timing was quite significant since it took place just two days after the signing the memorandum of understanding for TANAP with Azerbaijan. Putin hailed this agreement as a "New Year's gift" from Türkiye.[32] Türkiye also reached an agreement with Russia for the construction of its first nuclear power plant in Akkuyu, Mersin.[33] Moreover, the inauguration of the TurkStream project consisting of two pipelines delivering Russian natural gas to Türkiye through Black Sea in January 2020 further deepened Türkiye-Russia energy ties.[34] As clearly demonstrated by these examples, while enhancing its energy ties with Azerbaijan, Türkiye refrains from alienating Russia.

To enhance their role in the energy game, both Türkiye and Azerbaijan started to consider China as a possible energy partner as well. For the Azerbaijani side, there have not been many concrete steps, but the prospects are increasing. Chinese companies wanted to enter Azerbaijani energy market in the late 1990s and early 2000s; however, since most of the fields are occupied by the big Western energy companies, only some medium and small energy fields could be offered to Chinese companies by SOCAR. In recent years, the trade between Azerbaijan and China has significantly increased. In parallel with this development, there is a prospect of selling Azeri gas to Asia. Although it is not a pressing issue for either side for the time being, if such a project takes place Azerbaijan can find a new market, and China can diversify its suppliers. Thus, to enhance its energy ties with this region Asian Infrastructure Investment Bank approved a $600 million loan to TANAP in 2016.[35]

As for Türkiye, during Turkish President Gül's visit to China in 2009, the two countries signed a Memorandum of Understanding for the partnership in the energy sector, especially in renewable energy. As a result, Chinese investments in the Turkish renewable energy sector started to increase. Chinese companies made agreements with Turkish partners to initiate big solar energy projects and construct thermic power plants in recent years.[36]

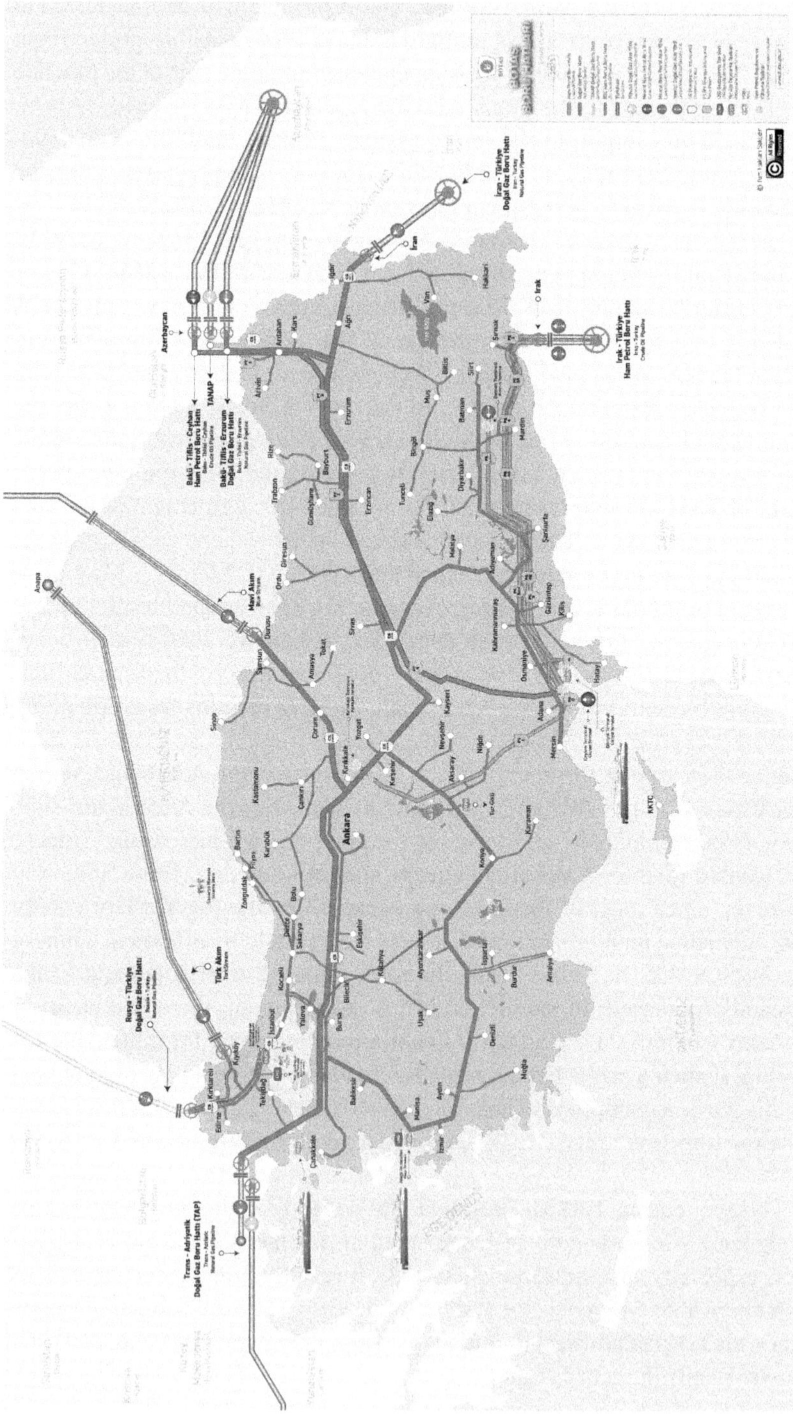

FIGURE 7.2 Natural Gas and Crude Oil Pipeline Map.

Source: Petroleum Pipeline Corporation (BOTAŞ), 2022.[30]

Since Türkiye desires to diversify its energy suppliers and aims to emerge as an energy corridor, Azerbaijan has a key role in Turkish energy policies. For the Azerbaijani side, selling its natural resources to Europe via Türkiye is a desirable option. Moreover, SOCAR aims to become an important player in the lucrative Turkish energy market as well. Thus, amiable political relations between Türkiye and Azerbaijan through intensifying energy ties creates synergy for both sides to reach their goals. Despite recent ups and downs in the relations, Azerbaijan and Türkiye often have complementary interests in the energy field. At the same time, Ankara has significant energy dependence on Russia, which causes it to play a constant balancing act with Kremlin. Moreover, in the past, while both countries entirely focused on Western routes and ties, connections with East Asia and China particularly with the scope of the Belt and Road Initiative (BRI), while still limited in scope, is receiving more attention.

As one of the cornerstones of human security, uninterrupted access to energy and freshwater resources emerges as an increasingly significant goal for states and individuals alike. Moreover, issues related with energy and water politics enable regional powers, such as Azerbaijan and Türkiye, to play a more active role in Eurasian politics, yet at the same time they create some vulnerabilities as well. Both countries are confronted with the challenging task of striking a delicate balance in the economy-environment-security triangle of the energy-water nexus.

Water security and politics have the potential to become the Achilles heel of Azerbaijan as a rising regional power in the medium and long term. Its high dependence on transboundary rivers causes a major challenge for Azerbaijan. The accessible freshwater quantity, and especially quality, also raises concern. As for Türkiye, due to its upstream riparian position for several river basins, and particularly for the Euphrates and the Tigris rivers, bestowing life upon Mesopotamia, there has been a debate about Türkiye's potential to become a "hydro superpower" in the Middle East and Eurasia. While launching the ambitious hydraulic based socio-economic development project, the Southeast Anatolia Project, at home, the transboundary nature of these river basins constituted important foreign policy challenges with Syria and Iraq, with some domestic implications on the Kurdish issue as well. Moreover, Türkiye also needs to deal with the challenges of pollution, salinization, and adverse effects of climate change in the future. Located in a semi-arid geography, the most urgent issue for Türkiye regarding climate change is water shortage which directly affects agricultural production and thus food security.[37] Per capita available water in Türkiye is only one-fifth of water-rich regions of the world such as North America and Western Europe. Moreover, 35% of Türkiye's resources are from transboundary rivers, which in turn necessitates the development of effective governance mechanisms and interstate arrangements and therefore resulting in reduced utilization of these resources.[38]

Finally, Azerbaijan and Türkiye have always placed the economy and energy dimension at the forefront in the energy-environment-security triangle of the energy-water nexus to reach their economic development goals and to enhance their position as significant regional actors. However, for a sustainable future, a more balanced approach emphasizing the pressing environmental needs is essential.

Energy-Water Nexus for Georgia: Hydroelectric Facilities, Rehabilitation, and Reconstruction

Georgia has been engaged in a sustained effort to expand its hydroelectric power generation capacity for many years. The Georgian government initiated numerous construction and rehabilitation projects. The plan is to provide sustainable (all year long) power to meet Georgia's growing demand as well as to increase its exports of electricity. Georgia is fortunate to have 26,000 rivers, constituting 60,000 km in total length, much of which traverses down from very mountainous terrain. Estimates suggest that at least 300 rivers are suitable for hydroelectric development. Financing remains a serious challenge, making Georgia dramatically dependent on environmental and social impact. Moreover, tense relations with the Russian Federation complicate some of the construction projects near Abkhazia in the Northern border regions. Financial support for hydroelectric plant development and related electrical grid modernization has come from the European Bank for Reconstruction and Development (EBRD), the Asian Development Bank (ADB), the World Bank, the United Nations Development Programme (UNDP), and the EU. There also have been substantial allocations from the US Agency for International Development (USAID), China, Japan, India, and the Swedish International Development Agency (SIDA).

The EBRD has a three-point strategy for its dealings in Georgia. First, it seeks to promote private sector investment. Second, it favors efforts to improve upon Georgia's capacity for use of renewable resources. And third, it tries to link Georgia into global and regional markets and value chains.[39] Hydroelectric power is critical in this strategy. By improving the hydroelectric capacity of Georgia, the EBRD can help to both improve its usage of sustainable energy and boost Georgian exports abroad.

Georgia is a regional favorite for the EBRD. It has classified Georgia to be a top regional reformer. The EBRD seems to prefer large-scale projects, often funding infrastructure and electrical renovation. A key example of this is the Black Sea Energy Transmission System. This project links Gardabani, Zestafoni, and Akhaltsikhe to ensure high-energy transfers. There is also a line from Alkhalsikhe to the town of Borçka in Türkiye. The System expands upon previously existing infrastructure in the region, and partners with EnergoTrans, a daughter company of the Georgian State Electrosystem, Georgia's state-owned electrical company EnergoTrans was designated responsible for the design and construction of the System. The ERBD approved a $120 million project to increase energy infrastructure link with Türkiye and has been considering rail projects in Tbilisi and port maintenance in Poti.

The Asian Development Bank's operations in Georgia began after the reforms of the Rose Revolution in 2003, but a strategy governing ADB projects in Georgia was only adopted in 2008. The focus has been on large-scale infrastructure projects in transportation, water systems, and energy production and transmission.[40]

The ADB assistance to Georgia in the energy sector reached a cumulative amount of 124.05 million dollars in 2020.[41] For instance, the ADB has taken the lead in Georgia's Regional Power Transmission Enhancement Project. This was a $48 million project approved in December of 2012. It involves the rehabilitation of eleven substations nationwide through the installation of digital controls and relay protection, promoting data acquisition, and efficient energy dispatching. The project was completed in 2016 with the construction of a new substation in Khorga, which aimed to ensure sales to Türkiye, fund management assistance, and conduct a study on hydropower.[42] The Georgian government anticipates these renovations to add $300 million annually to export earnings and to attract private investment. The goal is to increase sales to Armenia, Azerbaijan, Russia, Türkiye, and Central Europe. A key goal is enhanced trans-Caucasus integration of the energy market. The ADB has been collaborating with the EBRD, the World Bank, the EIB, the EU, and Germany's KtW in these efforts. These projects reflect the international dimension of energy politics in the region.

The ADB considered support of the Adjaristsqali Hydropower Project, to install two generators with a total capacity of 185 MW in southwestern Georgia. Electricity produced at this facility would be sold in the Turkish markets for nine months and serve the Georgian markets for the remaining three months each year. The project support was consistent with ADB's Strategy 2020 for inclusive and environmentally sustainable growth and regional integration.[43] World Bank involvement in Georgian hydroelectric projects is also significant. Much of it was organized under projects: (1) "Hydro Rehabilitation" and "Electricity Market Support" (for the installation of meters, hardware/software updates, better communication equipment, rehabilitation of substations, supply, and installation of computers to manage information, internal management aid, consultation for engineering, and implementation); (2) "Energy Transit Institution Building" (committing over 75% of the cost, $9.63 million of $12.29 million, for consulting, legal advice, and risk mitigation, and financial audits); and (3) "Energy Sector Adjustment Credit Project" (operating between 1999 and 2002 to help the new Georgian government privatize and effectively regulate the energy sector, with a special focus on oil and gas committing 100% of the $25 million cost).[44] In 2015, the World Bank Board of Executive Directors approved an IBRD loan of $40 million for the Georgia National Innovation Ecosystem (GENIE) project.[45]

The Georgian government tends to prioritize privatization of the energy sector, for example, in the creation of the Electricity Wholesale Market and the National Energy Regulatory Commission. However, there are remaining problems about tax collection and related financial impact, and the lack of control over energy suppliers. The Georgian Government has expressed concern at the disproportionate supply coming out of the Enguri Dam. Thus, it has prioritized general rehabilitation of hydropower facilities, noting that many projects were underway that would greatly increase capacity. Rehabilitation programs are often financed through bilateral and multilateral investment projects.

A 2013 ADB assessment report praised Georgia for the reforms it had already undergone, specifically the creation of the National Energy Regulatory Commission and the unbundling and privatization of many factors of energy production, transmission, and distribution. The experts also expressed concern that the Enguri, Vardnili, and Khrami hydropower plants remained in state hands.[46] Collection of billed electricity has improved to 75%, with most regions enjoying 24-hour access to power. Transmission losses were down from 16% in 1995 to 2% in 2012, and Georgia has transformed into a net exporter of electricity.[47]

There is progress to be made in expanding private participation in the energy sector. To accomplish this, the Georgian government will need to take major steps in building a legal, institutional, and regulatory framework. Specifically, there is no clear environmental regulation, insufficient private-public partnership framework, lack of transmission capacity, and unclear off-take agreements, all of which constrain private sector development. The ADB stressed that these issues are stalling construction of many private power generation facilities. Additionally, there are inadequate market rules for settling and balancing accounts across borders, poor trans-state transmission infrastructure (even including Black Sea Transmission Line), and potential for institutional conflict of interest between SOEs, all of which hampers international trade. The Ministry of Energy and Natural Resources lacks capacity to develop and implement efficient energy policy, and there is neither a construction code nor efficiency or labeling standards.

The ADB report recognized that the Georgian government understands the problems it faces. The government is in the process of harmonizing technical and legal aspects of Georgia's energy system with neighboring states to boost exports, specifically Türkiye. Moreover, work is underway to improve efficiency, domestic production, promote financial and technical sustainability, and further the privatization project. Nevertheless, the ADB believes that it has a role to play in improving the Georgian economy, especially in improving efficient energy production and usage and enabling financially and environmentally sustainable private sector involvement in hydropower plant development.

There is substantial support from international institutions. The United Nations Development Program reported on April 18, 2011 support for a hydropower project of 15,000 megawatts. This is in conjunction with German KfW funding for a small HPP project, together with 5 million euros to the Georgia Renewable Energy Fund. Georgian banks TBC and VTB received preferential low interest loans from the Renewable Energy Fund at the Ministry of Regional Development and Infrastructure of Georgia. Additionally, relevant USAID Projects include: (1) Georgian Energy and Water Regulatory support, September 2010–September 2016. This is a partnership with the National Association of Regulatory Utility Commission; (2) Black Sea Transmission Planning effort, September 2010–September 2016 in partnership with the US Energy Association to examine a regional high-voltage transmission network to see where investments can improve efficiency and capacity; (3) Power and Gas Infrastructure Program, May 2010–November 2013, partnering with the Government of Georgia to set strategy at the national level for Infrastructure Oversight and Capacity Building.[48]

Specific cases of hydroelectric plant development projects include the following:

Enguri

The Enguri dam is the crown jewel of Georgian hydroelectricity. Constructed by the Soviets, it had fallen into serious disrepair during the last years of the Soviet Union and the initial years of the Georgian transition. Hydro water potential of the river Enguri is approximately 3,500 megawatts annually and the electricity generation potential is 10.3 billion kilowatt-hours.[49] The World Bank and the EBRD helped to fund its renovation, with the first phase beginning in 1998. According to the EBRD, this phase was co-financed by a $10 million grant from the European Union. The second phase of the project focused on supporting civil works and the other two units of the dam. It was co-financed by another $2 million grant from the European Union. The total cost for this renovation was projected to be $48.5 million in 2007, though a 2013 estimate would put the bill as high as $70 million. These financing arrangements were conditional on Georgian legal reforms to improve the efficiency and sustainability of the energy sector. The debt is to be repaid over the course of 15 years. SinoHydro of China helped in the construction, and China contributed to the financing. The European Investment Bank also increased its support by 3.5 million Euros to finance the completion of rehabilitation works at Enguri and Vardnili PPP.[50]

Enguri itself is situated in Svaneti, on the border with Abkhazia. The renovation began shortly after the conclusion of the civil war between the Abkhaz and Georgians, and reports on Enguri as late as 2011 emphasize the dam's role as a stabilizing force in the region. It's noteworthy that while the 2008 war with Russia was nominally predicated on securing Abkhaz independence and that the Abkhaz have since enjoyed a nominal independence, Enguri has never been damaged in an armed attack. But there were mutually exclusive claims offered frequently as to whether it is rightly owned by the Abkhaz or the Georgians.[51]

The renovations of Enguri yielded quite positive results. In 2007, the EBRD thought that their renovations would improve capacity by 15% and provide up to half of Georgian energy; a 2011 report indicated that production had been improved by 30%, and by 2013 Enguri was providing 40% of Georgian electricity. Furthermore, the Enguri project has allowed Georgia to join the Kyoto Protocol's Clean Development Mechanism. Enguri is projected to produce more than 5.8 million carbon credits over the ten-year crediting period—credits which can be sold abroad. The policymakers also view Enguri as central to the future of Georgia's electricity network. In 2012, the ADB issued a $48 million loan to Georgia to construct a power system guaranteeing exports from Enguri to Türkiye via Poti.

In the tense months after the conflict between Russia and Georgia in 2008, Georgia announced a deal with the Russian energy company RAO U.E.S. that would permit Russians to buy electricity from Enguri for distribution in Abkhazia and for export to Türkiye via Georgia. This raised some concern that the Georgian government would lose its control over Enguri, given its proximity to Abkhazia, firmly protected by Russia. The Georgian government noted that Enguri was often the site of border incidents, citing one on 14 January 2009 in which unidentified assailants shot at police posts in the region from the Abkhaz side for twenty minutes. Reportedly, police did not return fire, and there were no casualties. The deal would seem to recognize Tbilisi's sole jurisdiction over Enguri and would rectify the prevailing situation that had Abkhazia consuming 25% of the Enguri power output without payment to Georgia.[52] The Georgian Energy Minister, Khetaguri, said on January 13 that:

> Even if, let's say, today the Russians switch off the generators, there is nothing we can do about it, but with this deal we gain additional guarantees that this will not happen, because the Russians themselves will be interested in the full functioning of the plant. The Russian company also guarantees its commitment through its serious assets in Georgia.[53]

Ultimately, according to Georgia's Energy and Water Supply Regulatory Commission's report, by 2015 the Enguri renovations enabled Georgia to distribute 1,797 million kWh electricity to Abkhazia and to provide 17.31% of Georgia's overall consumption.[54]

Khrami

The Khrami facilities are less massive than their Enguri counterpart but are still relevant to the Georgian economy and energy strategy. The InterRao Khrami I & II are a $204 million investment. They produce 227.2 MW, and are sponsored by Inter RAO U.E.S. It is noteworthy that InterRAO U.E.S. is under the majority-stake holdership of the Russian government. Additionally, InterRAO U.E.S. owns and operates the Mtkvari Thermal Power Plant, bringing its total production capacity in Georgia to 827.2 MW. This is not a massive amount of production capacity, but it is indirectly owned by the Russian government. The Khrami facilities are part of the larger push towards exporting electricity to Türkiye. The World Bank categorizes Khrami as a "cross-border" hydroelectric plant.

Khudoni

The Khudoni Hydro Power Plant is the second largest project in Georgia after Enguri.[55] The World Bank conducted a Pre-Investment Facility for Khudoni, specifically focusing on roads and infrastructural improvements. The Saakashvili government worked seriously on implementation, but the project elicited substantial domestic opposition. The general line of argument was that large-scale dams would harm Georgia's environment without boosting Georgia's sluggish economy. Georgian law did not then require environmental or social impact assessments, and those required by international funding agencies were reportedly kept secret by the Saakashvili government. Moreover, potential investors were incentivized with tax breaks and other favorable terms to encourage electricity exports from the project.

Located in the Svaneti region, near Abkhazia, the Khudoni project would disrupt the homes and lives of thousands of Svans in a high mountain region designated by UNESCO for special recognition and preservation. This initiative constituted a major human security challenge at the local level. Protestors also expressed concern about the potential for seismic vulnerability; indeed, Georgian green party officials and environmental activists warned of the potential for flooding substantial parts of Western Georgia in the event of an earthquake.

Soviet plans for hydroelectric generation in this region were shelved due to protest, while independent Georgia under Saakashvili seemed not to be dissuaded. The project was to begin in 2012, with generation operations expected by 2018. Georgian Dream politicians promised an end to large-scale hydroelectric projects, with special reference to Khudoni.[56] The reality of national energy requirements, in the absence of fossil fuel sources, however, remain a challenge, especially given the initial success in exporting electricity from the completed construction and reconstruction to Europe and Eurasia—at a time when other prospects for export income remain scant. Ivanishvili stated his preference for smaller scale hydroelectric projects and solar power development.[57]

A petition signed by 26 NGOs and posted by the Georgian Young Lawyers' Association October 17, 2013, argued that the construction of Khudoni is illegal. The petition highlighted that the appropriate permits were never issued, and the office of the Prime Minister hijacked the process. The signatories argued for return of the decision process regarding Khudoni construction to a legal framework, an end to the external pressure and lobbying to shove construction forward, and an effort to eliminate elite interference and conflicts of interest in the HPP decision process. It also criticized the ruling Georgian Dream government for backtracking on promises to end large HPP projects, noting that nothing in the economic circumstances of Georgia would necessitate that reversal in policy.[58] Construction was in effect halted by Greens and Nationalists, restarted by Saakashvili, and stopped again by Georgian Dream.

Nonetheless, the Georgian Dream administration moved to the construction phase, despite campaign promises in opposition to the posture of the Saakashvili administration. Fissures developed among the coalition members. Some former members have joined the continuing protests against the Khudoni project. An Indian firm, Trans Electric, was engaged in the construction. The State Bank of India was interested in investing in the Khudoni HPP and met with the Georgian Ministry of Energy and Natural Resources and TransElectric, Khudoni's operating company. The State Bank of Georgia has been financing 70% of construction. Khudoni has an installed capacity of 702 MW and an annual output of 1,539 billion KWh. The planned construction project was to install a 200-meter-tall arch dam for $1.125 billion.[59] Recently, the Government of Georgia is negotiating with Transelectrica to cease the Khudoni Hydropower Plant agreement, which was signed on April 28, 2011 as Transelectrica cannot carry out the project.[60] After the completion of negotiations, the government is planning to implement the project itself.[58,61]

Smaller Scale HPPs

The European Bank for Reconstruction and Development funded two hydroelectric plants on the Namakhvani river with a ten-year $500 million Eurobond at the London Stock Exchange. This initiative was aimed at boosting Georgian energy exports, as well as improving the transparency and governance.

Other developing hydroelectric facilities in Georgia include the Ladzhanuri Hydro Power Plant, the Samtskhe-Javakheti Hydro Power Plant, the Tvishi Hydro Power Plant, the Vardnili Hydroelectric Power Station, the Zhinvali Dam, and the Zhoneti Hydro Power Plant. Georgian Dream officials expressed some concern about planned HPP construction on the Ajaristskali River, citing possible negative environmental and social impacts.[62] After an assessment by the Georgian Energy Ministry in mid-2013, the new Government was reassured.[63] In October 2013, the Ministry of Energy signed contracts with GHP to build two additional hydroelectric stations: the UbisaHES, which will have a capacity of 4 megawatts, built in Imereti; and the AkvaretaHES, which will have a capacity of 18 megawatts, and be built in Ajara. The total cost is estimated at $30–35 million, and the project engages Chinese firms that see these as pilot projects for future work.

As indicated by these numerous projects, water politics have a decisive impact for Georgia from the local to the national level, as well as from the regional to the international level by effectively linking energy, environment, and geopolitics.

Energy and Environment in Türkiye and Azerbaijan

The concept of sustainable development gained more importance in the last 20 years. Among the UN Sustainable Development Goals (SDGs), the need for ensuring everyone's access to affordable and clean energy is highlighted.[64] However, there must be a delicate balance between energy choices and environmental impact. The energy sector presents significant environmental challenges in Azerbaijan and in Türkiye with increasing pollution caused by energy production and consumption.

The Turkish Ministry of Energy has been focusing on increasing renewables for diversification. The Turkish Parliament passed the Law on the Utilization of Renewable Energy Resources in 2005. Subsequently, the General Directorate of

Renewable Energy replaced the sub-unit of the Ministry of Energy called the Electrical Power Resources Survey and Development Administration (EIE). After the definition of the legal basis, renewable energy investments started to increase, particularly on wind power and geothermal energy. There is a rising demand from various sectors in Türkiye, including the public sector, some private investors, and environmental lobbies, for better and more transparent regulation of the carbon market. Accordingly, electricity generation was 304.3 TWh in Türkiye in 2019, including 37.2% coal, 29.2% hydro, 18.7% natural gas, 7.2% wind, 3.5% solar, 2.9% geothermal, 1.1% bioenergy and waste, which shows 56% increase for electricity generation since 2009.[65] There was also an initiative to develop a national action Plan on Energy Efficiency in Collaboration with EBRD.[66]

Another problematic issue for the Turkish energy sector is the use of coal, due to its significant share in Türkiye's energy mix and adverse climate implications (see Figure 7.3). Given Türkiye's emphasis on utilizing local resources of coal to enhance its energy security, an important way to increase efficiency and to mitigate detrimental environmental effects could be to promote and implement clean coal technologies. While still being a less environmentally friendly option than natural gas and renewables, these technologies decrease the CO_2 emissions from the burning of coal and increase efficiency obtained from each ton of coal.[67] Clean coal technologies are not widely used in Türkiye yet. However, the Ministry of Energy encourages their usage and supports research and development projects related with them.[68]

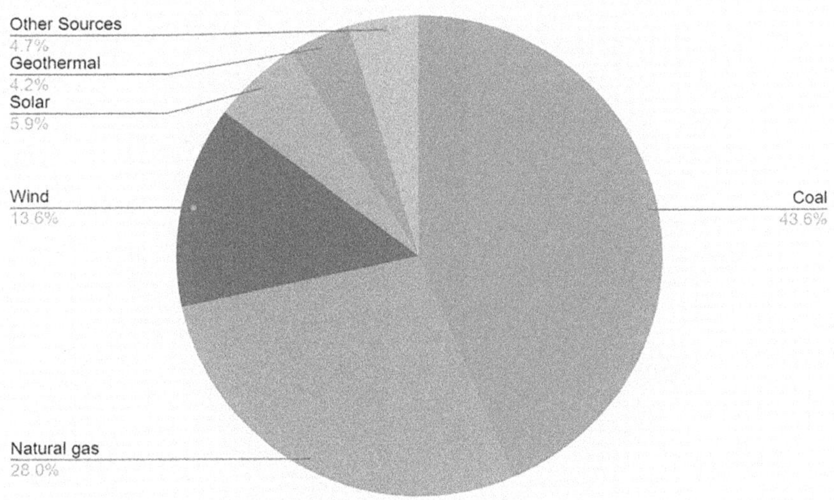

FIGURE 7.3 The Share of Energy Resources in Türkiye's Total Electricity Supply (2022).

Source: Republic of Türkiye Ministry of Energy and Natural Resources, 2022.[69]

As a major producer of fossil fuels, almost 98% percent of the Azerbaijani energy consumption is based on natural gas, petroleum, and other liquids (see Figure 7.4). In comparison with the Turkish consumption patterns, the limited usage of coal combined with the high use of natural gas in Azerbaijan is an advantage in terms of carbon emissions. In this context, Azerbaijani ratio of CO_2 emissions to TPES is lower than the world average, and the ratio of emissions to GDP is higher than both Türkiye's score and the world average.[70]

In general, Azerbaijan faces serious environmental challenges due to energy production. Caspian petroleum and petrochemical industries have contributed to the water and air pollution in Azerbaijan. The Azeri government started some initiatives to cope with these challenges. An important step was the state program on the "Use of Alternative and Renewable Energy," which was approved in 2004. The program mainly focuses on wind and hydro projects and tries to attract investments to these fields. However, only few activities have taken place in these areas since then. The use of renewables only increased from 1% to 2% between 2000 and 2010. Nevertheless, the Azerbaijani State pursued renewable energy projects as they launched a program with UNDP in 2011.[72] The renewable energy targets

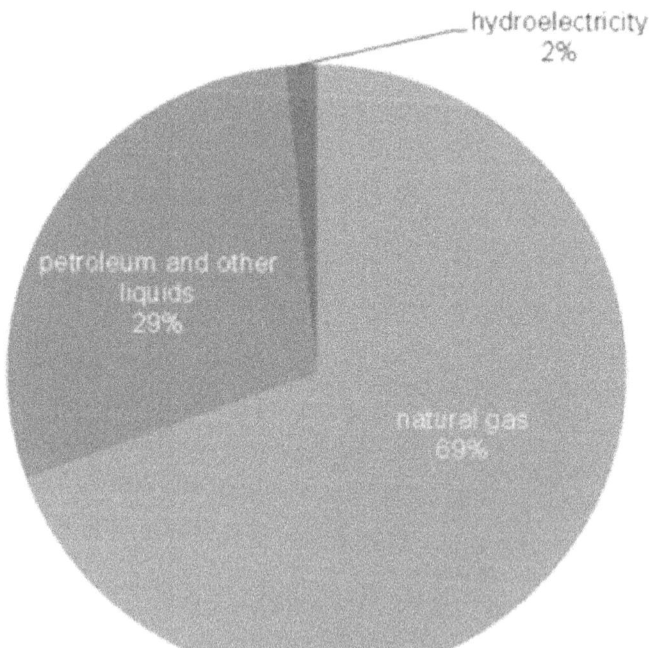

FIGURE 7.4 The Share of Energy Sources in Azerbaijan Total Energy Consumption (2020).

Source: EIA based on data from BP Statistical Review of World Energy, 2021.[71]

for Azerbaijan envisioned 2000 MW of installed renewable energy supply capacity by 2020.[73] However, installed capacity on renewable energy sources (large hydropower plants are excluded) was 168,3 MW in 2020 which made up 2.2% of total electricity generation capacity.[74]

The perceptions of energy efficiency vary between developed and developing countries. Most developed countries pay more attention to energy efficiency in comparison with developing countries. In the developing countries, the measures for efficiency and environment are usually neglected, since they increase the costs.[75] This is why developed countries are often the forerunners of application and the development of energy efficiency policies in the world.

Energy intensity is used as the indicator that shows the countries' energy efficiency performances in general. Energy intensity shows the amount of energy that is used for one unit of GDP.[76] There is a negative correlation between energy intensity and energy efficiency. Energy intensity indicator is commonly used for monitoring and comparing the energy efficiency performances of the countries.

Türkiye has an average performance in terms of energy efficiency. Considering the increasing energy needs and future goals, it is important for Türkiye to have a better energy efficiency performance. Although Türkiye's energy efficiency performance is better than the world average, the energy intensity score of Türkiye is relatively higher than the OECD countries.[77] According to International Energy Agency (IEA), Türkiye's energy intensity is 0.12, while this value is 0.11 for the OECD countries and 0.18 for the world average.[78] Although there is a decrease in Türkiye's energy intensity value, it has a slow pace. In Türkiye, in parallel to rising energy needs and energy expenses, the emphasis on energy efficiency has increased since 2000. In this period, the legal basis for these measures was prepared. As a critical step, the government passed the Energy Efficiency Law in April 2007. Following its enactment, secondary regulations were put into action in 2007 and 2008.[79] This law defined the main regulations and policies for increasing energy efficiency in houses, industry, and transportation during the phases of production, transfer, distribution, and consumption.[80,77] The government published the *Energy Efficiency Strategy Paper* in February 2012 in order to create an energy efficiency vision for Türkiye for the period of 2012–2023.[81]

In the framework of the stated laws and regulations, there is a serious potential for energy saving. While Türkiye's energy efficiency is relatively low in comparison to the other OECD countries, in comparison to BRICS (Brazil, Russia, India, China, South Africa), Türkiye has an impressive energy efficiency performance. Thus, through prudent policies, Türkiye could target becoming a leading country among the developing countries in energy efficiency and reaching the OECD average. In this context, *Energy Efficiency Strategy Paper* states that Türkiye aims to reduce its energy intensity by about 20% until 2023, the 100th anniversary of the foundation of the Republic of Türkiye.

Azerbaijan's energy efficiency performance is quite dismal. According to IEA, Azerbaijan's energy intensity value is 0.24, which is higher than the world average.[82] The main reason behind this performance is the rich energy resources of the country. Since the accessibility of energy is not a problematic issue for the energy-rich countries, they pay less attention to energy efficiency. In that respect, relatively poor performance of Azerbaijan in energy efficiency is not surprising. To better assess Azerbaijani energy efficiency performance, it is important to compare it with other energy-rich countries. When we examine the energy producers of the Caucasus and the Middle East, we can observe that Azerbaijan's score of energy intensity is not that high. Most of the regional energy producers such as Russia (0.41), Iraq (0.26), Iran (0.51), and Saudi Arabia (0.33), however, have even worse performances than Azerbaijan.[83]

There is no separate law in Azerbaijan concerning renewable energy and energy efficiency. However, the *Law on the Usage of Energy Resources*, which passed in 1996 defines the legal background for the policies for the use of energy resources.[84] Also, the State Program on the Use of Alternative and New Energy Resources was prepared in 2004. The Azeri government intensified its efforts concerning renewables since 2011 with the preparation of a national strategy for the use of alternative and renewable energy sources (RES) for 2012–2020 and a RES law has been under development. The Azeri government has committed 60 million dollars for the development of renewable energy resources in Azerbaijan including hybrid (solar, wind, and biomass) RES facilities.[85] The progress in terms of the implementation of actual projects, however, has been rather limited.

So far, in the energy realm on the economy-environment-security triangle, economic and security issues have been at the forefront at the expense of environmental concerns. On the energy-water nexus, while there are some promising initiatives a similar picture emerges.

Geopolitical Factors in Action: Emerging Water Constraints and Conflicts in the South Caucasus

The water politics of Azerbaijan is a hotly debated issue due to political disputes between Azerbaijan and Armenia over Nagorno-Karabakh and its implications for hydropolitics. The rapid pace of industrialization, increasing usage of water resources, and low levels of cooperation for the ecological security of transboundary water systems constitute additional challenges.

The main water sources of Azerbaijan are the Kura and Aras rivers, both of which originate in Northeastern Türkiye. The Kura River flows through Georgia and Azerbaijan, whereas the Aras River flows through Georgia, Armenia, and Azerbaijan, respectively. The Qanykh River, which originates in Georgia, is the third most important water source in Azerbaijan in terms of volume. The local flow volumes of Kura River, Aras River, and Qanykh River are between 5 and 6 km^3, 1.6 and 1.8 km^3, and 1.4 and 1.6 km^3, respectively.[86] The Kura River originates in Kars in northeast Türkiye, then flows into Georgia and flows into Azerbaijan in the northwest. The total length of the Kura River system is 1,515 km, of which 900 km is located within Azerbaijan. The Aras River also originates in the northeast of Türkiye and forms the border between Türkiye and Armenia, Türkiye and Azerbaijan, Iran, and Azerbaijan, before flowing into the eastern part of Azerbaijan. About 100 km downstream of the border, it joins the Kura River, which continues to flow southeast towards the Caspian Sea. There are also two relatively large water reservoirs in Azerbaijan, Mingechevir and Shamkir with a capacity of 16.07 km^3 and 2.68 km^3 respectively.[87] It is also noteworthy that the territories over which Azerbaijan regained control in the Second Karabakh War in 2020 include an important water resource. Situated in Nagorno-Karabakh with a capacity of 0.56 km^3, the Sarsang reservoir was controlled by Armenia since 1991 and Azerbaijan was denied access to water in the desperately needed Tartar region of Azerbaijan. As the fifth largest water reservoir in the country, it also contributes to diminishing the water shortage in Azerbaijan to some extent.

There are 8,359 rivers of various lengths within Azerbaijan; however, 8,188 of these rivers are less than 25 kilometers in length, and only 24 rivers are over 100 kilometers long. The geographical distribution of the rivers can be categorized in three groups within the country: the Kura basin rivers, the Aras basin rivers, and the rivers that flow directly into the Caspian Sea. Water resources are distributed irregularly over administrative districts with Absheron and Kura-Araz lowlands being the most water-poor regions.[88]

Since only 25–30% of Azerbaijan's rivers have domestic origins, the transboundary nature of its major water resources makes hydropolitics an integral part of Azerbaijani foreign policy. Azerbaijan, which relies extensively on water flow from other countries, stresses that it is the only Southern Caucasus country to sign the Convention on the Protection and Use of Transboundary Watercourses and International Lakes (Helsinki Convention).[89] In addition to the lack of effective governance mechanisms regarding the transboundary river systems, high levels of pollution is also a significant problem for Azerbaijan. Due to the unsatisfactory performance and/or lack of treatment facilities, Georgia, Armenia, and Azerbaijan often discharge polluted waste waters into the Kura River and its tributaries.[90] Thus, large concentrations of heavy metals that exceed allowable limits can be found in the water sources. Moreover, a 2009 UN report indicated that 76% of the Azerbaijanis in urban cities had access to piped water, whereas only 19% of the population in rural areas had access to piped water.[91]

There has been some important progress in this realm. According to the 2015 UN-Water Global Analysis and Assessment of Sanitation and Drinking-Water:[92]

> In recent years, 4,280 kilometers of water supply lines have been laid; and 36 water inlet units, 164 reservoirs and 74 pumping stations have been built. Also 175 artesian and sub-artesian wells were commissioned. Over the past 10 years, 20 sewage pumping stations and three wastewater treatment plants have been built; three wastewater treatment plants were repaired; and 1490 miles of sewer lines were improved.

Erosion is another major issue that affects almost 43% of the country. According to United Nations Economic Commission for Europe (UNECE), the major problems affecting irrigation have been the deterioration of infrastructure and pumping equipment due to insufficient maintenance; heavy reliance on pumped irrigation; and inefficient water distribution and application.[93] UNECE indicates that 40–50% of the irrigation infrastructure is in dire need of renovation. Thus, the inefficient use of water and water losses in irrigation constitute major problems for water resources and soil.[94]

A cross-country comparison of water sources indicates that the situation for Türkiye is much better. Taking their land surface into consideration, Türkiye and Greece have 1.77%, and 0.99% of water, respectively.[95] More importantly, the dependency ratio for renewable water sources for Türkiye, Greece, Bulgaria, Syria, Iraq, Georgia, Armenia, Azerbaijan, and Iran are 1.5%, 15.2%, 1.4%, 72.4%, 60.8%, 8.2%, 11.7%, 76.6%, and 6.8%, respectively.[96] These findings indicate that Türkiye is the most self-sufficient country in terms of water in its region, with Syria, Iraq, and Azerbaijan constituting the most externally dependent countries of the world.

There are several institutions involved in water management in Azerbaijan. The most important authority is the Ministry of Ecology and Natural Resources, which has overall responsibility for water resources and for the prevention of pollution. There is also the Committee on Ecology and the Nature Use, which is in the charge of monitoring salinization and water pollution. The Ministry of Health and its Center for Epidemiology and Hygiene unit is responsible for monitoring drinking water quality. The State Committee on Amelioration and Water Management is responsible for monitoring water use and for issuing permits for surface water.

On the foreign policy front, the implications of the Nagorno-Karabakh conflict on water policies are quite serious for Azerbaijan. The Armenians and the Azerbaijanis shared the water in the Karabakh region under the USSR, but they have been under Armenian control since the early 1990s. The Armenians view the control over water resources as a part of their sovereign right, regardless of how their water policy affects the Azerbaijanis downstream. Azerbaijan had been accustomed to access under the Soviets and views the Armenian water policy as deliberately hostile.[97] The journalist Samira Ahmedbeyli indicates that "[life has become difficult for] the villagers on the Azerbaijani side of the line because much of their water supply had been cut off as a result of the conflict." Some Azerbaijanis call this problem "an ecological threat" to the future of their country and of the region. In fact, Karabakh region has high quality water, as residents of the Karvachar region note: "We have the most delicious and cleanest water. It does not need filtering. You can drink it straight from the river."[98] Another feature of the region that results in further political fragility is that most water sources in the former Autonomous Region of Nagorno-Karabakh have their origin outside of its administrative limits.

According to an expert of the region, David Babayan, the rivers Terter and Khachen bring in 83.4% of the yearly average of Karabakh's main water supply.[99] After occupying Terter Plain, Armenia had stopped allowing water flow to the Plain since 1991; however, after Second Nagorno Karabagh war, the Azerbaijani army regained control over Suqovuşan and the Terter River started to flow once again in the region.[100]

As for Türkiye, political disputes emerging out of water politics between Ankara and its neighbors are often overshadowed by other broader problems, such as PKK terrorism. However, there is also a major problem because of the uneven water distribution in regions surrounding Türkiye.

The fresh water sources of Türkiye are diverse: Türkiye is divided into 26 hydrological basins with large differences in specific discharge. Most rivers originate in Türkiye and there are more than 120 natural lakes and 579 artificial lakes. General Directorate of State Hydraulic Works in Türkiye, DSI, acknowledged that Türkiye is significantly rich in terms of ground reserves with an annual 11.43 km^3 of water.[101] Recent figures show that "Türkiye's total usable water potential is 112BCM, of which 98BCM is surface water and 14BCM is groundwater."[102]

In terms of the transboundary rivers, most important of which are the Euphrates and the Tigris, Türkiye contributes to about 90% of the total annual flow of the Euphrates, 38% directly to the main Tigris river, and another 11% to its tributaries joining the main river further downstream in Iraq. In general, the streams and rivers of Türkiye have major seasonal and annual flow fluctuations. Türkiye shares many of its rivers with other countries: about 615 km, or one-fifth of the total border length of 2,950 km between Türkiye and other countries is formed by rivers. 238 km of the borders with Bulgaria and Greece, 243 km with Armenia and Georgia, 76 km with Syria, and 58 km with Iraq and Iraq consist of water.[103] This situation resulted in the signing of treaties with some of its neighbors to control the management of these rivers, such as the treaty between Türkiye and USSR in 1927 and the one between Türkiye and Greece in 1973.[104] Similarly, the 1946, 1980, and 1982 treaties between Türkiye, Iraq, and later Syria regulate the amount of the flow of the Euphrates and the Tigris from Türkiye to other signatories. Türkiye also unilaterally guaranteed that it will allow 500 m^3/s (15.75 km^3/year) water flow across the border to Syria, but no formal agreement has been obtained so far on the sharing of the Euphrates River.[105]

Access to water supplies and sanitation in Türkiye is relatively high. Based on household surveys and census results, the Joint Monitoring Programme for Water Supply and Sanitation estimates that 100% of the Turkish urban population had access to an improved water source.[106] According to World Bank data, 85% of all population has access to improved sanitation facilities while in urban areas, the number reaches up to 98%. In rural areas, 89% is using basic sanitation facilities.[107] In 2020, access to improved sanitation facilities was 100% for the urban population while it was 91.6% for the rural which made 97.3% of the total population.[108]

The DSI and the GDRS (General Directorate of Rural Services) construct the drainage infrastructure in the irrigation schemes. The main, secondary, and tertiary drainage canals are constructed by the DSI, while the GDRS builds the on-farm drainage systems. For the operation, maintenance and repair of drainage canals, 38,278 km of service-road have been built by the DSI.[109] DSI started irrigation modernization in 2008 which

> allows the use of high efficiency on-farm irrigation systems such as drip and sprinklers and increases the net irrigation area under operation by almost 30 percent, from 2.91 m ha to 3.75 m ha; raising irrigation efficiency from 42 to 50 percent; expanding the use of water saving modern irrigation systems by 10 percent per year; and decreasing the use of groundwater by 5 percent during the plan period.[110]

During recent decades, the GDRS has carried out many small on-farm water development works, for example, on-farm drainage systems and saline and alkaline soil reclamation. The area protected from flooding amounts to almost 397,302 ha. It was estimated in 1992, that of the total area operated by the DSI, about 41,000 ha was salinized by irrigation. In 2004, the total area salinized by irrigation in Türkiye was estimated at 1.5 million ha. An area of 2.8 million ha is affected by waterlogging and drainage problems.[111] In addition to these, traditional irrigation, such as surface irrigation, results in relatively more inefficient use of water due to leakages, evaporation, and operational losses. Approximately 81% of Türkiye's total irrigated area is through surface irrigation and 19% is using more efficient pressurized irrigation methods.[112] With climate change and potential increase in the duration, frequency, and intensity of flooding and droughts, these issues will be particularly significant.

Türkiye and the Greater Middle East: Challenges of the Water-Energy-Food Nexus

The Middle East region is the world's richest region in terms of energy reserves and the poorest one regarding water resources. The Middle East region has abundant petroleum fields and 32% of the world's known natural gas reserves are in the region.[113] Middle East produces 37% of the world's oil and 18% of its gas. while it has 65% of global oil reserves and 45% of natural gas reserves.[114] Moreover, increasing the share of renewable energy in their energy mix, Middle Eastern countries have spent "$150bn in solar power, and $28bn on wind, waste-to-energy, hydro and geothermal power plants" for the renewable projects that mostly cover the oil-producing Gulf.[115] In contrast to its abundant energy resources, the Middle East region is the world's "most water-scarce region" which is even worsening with the armed conflicts, climate change, and economic decline.[116] The water crisis causes declines in water resources per capita which is already one-sixth of the global average.[117] Additionally, it is expected that in 30 years, groundwater will run out in most parts of the region.[118]

Water constitutes an increasingly important dimension of regional politics with the continuing growth of population and decreasing freshwater sources; experts indicate that the quantity of water available to everyone is predicted to decrease by 30–40%.[119] Water has a significant role in Turkish politics as well. In this context, the political disputes between Türkiye and its southeastern neighbors Syria and Iraq, and to a lesser extent Georgia and Azerbaijan resulting from water have an impact on Türkiye's relations with those countries and even the EU. In relation to hydropolitics, we will be zooming on the Tigris-Euphrates basin within the context of the greater Middle East as it complements our discussion regarding Eurasia.

Türkiye is in a strategically strong position in the Middle East being the only country that enjoys abundant ground and surface water resources. In contrast, Syria depends heavily on the waters of the Euphrates, whereas Iraq is reliant upon the Euphrates, but also uses the Tigris River as an alternative source of water. Some Turkish diplomats suggest that Türkiye believes that it has the right to use water as it wishes, similar to the case of Arab governments being the only power controlling the oil resources in their countries.[120] In this regard, serious political problems did not emerge until 1960, when the midstream and upstream countries, Syria and Türkiye respectively, initiated projects for developing the waters of the Euphrates. The most extensive and ambitious among these projects is the Southeast Anatolian Project that involves both the Euphrates and the Tigris. Syria and Iraq have been strongly opposing this project, because they fear that Türkiye's consumptive use of the Euphrates waters will affect the quality and the quantity of flow downstream. The dispute on the project between the three countries has been going on for decades and brought Türkiye and Syria on the brink of war in 1998.

The construction of the Ataturk Dam, completed in 1992, has been widely portrayed in the Arab media as a belligerent act, since Türkiye began the process of filling the Ataturk dam by shutting off the river flow for a month.[121] Both Syria and Iraq accused Türkiye of not informing them about the cut-off, thereby causing considerable harm. Iraq even threatened to bomb the Euphrates dams. Türkiye responded by arguing that its co-riparians had been informed in a timely manner that river flow would be interrupted for a period of one month for reasons of "technical necessity."[122] Türkiye returned to previous flow-sharing agreements after the dam became operational, but the conflicts were never fully resolved as downstream demands had also increased in the meantime.

Several crises have occurred in the Euphrates-Tigris basin, due to lack of communication, conflicting approaches, unilateral development by all sides, and inefficient water management practices. Downstream countries have accused Türkiye of violating international water laws about the Euphrates and the Tigris rivers. Iraq and Syria consider these rivers as international and thus claim a share of their waters. Türkiye, in contrast, focuses on the concept of rational utilization of transboundary waters. Türkiye also voted against the United Nations Convention on the Law of Non-navigational Uses of International Watercourses. According to Türkiye, if signed, the law would give the lower riparians a right of veto over Türkiye's development plans. Consequently, Türkiye maintains the argument that the Convention does not apply to them, and it is not legally binding.[123] In April 2008, Türkiye, Syria, and Iraq decided to cooperate on water issues by establishing a water institute consisting of 18 water experts from each country to work toward the resolution of water-related problems between the three countries. This institute was to conduct its studies at the facilities of the Ataturk Dam, the biggest dam in Türkiye. They planned to develop projects for the fair and effective use of transboundary water resources.[124] However, given the troubled relations between Türkiye and Syria with the Arab spring and the current ongoing conflict within Syria, such cooperation seems highly unlikely in the short and medium term.

The Syrian war caused destruction of cultivated lands along the western coast and eastward along the northern border with Türkiye, as well as the contamination of land and soil by war material which intensified the erosion process and caused a loss of agricultural production.[125] As a consequence, soil erosion caused both loss of fertile land and increased sedimentation and pollution in the rivers and streams which reduced fish production by 63%.[126] As a result of the military campaigns of Türkiye with the allied Syrian militias to northern Syria, control of Afrin, Ras al-Ayn, and Tel Abyad is taken from the SDF fighters.[127] Recently, declines in water levels on the Euphrates River caused by climate change caused tensions between Türkiye and Syrian Kurdish forces as well.[128] The Syrian refugee problem constitutes a key challenge for the EU, as well as for Türkiye which hosts more than 3.6 million refugees. In this context, it is noteworthy that the Syrian refugees are not only fleeing civil war, but they are also forced out of their lands, in important respects, because of water constraints on rural agricultural populations aggravated by military conflict.

Turkish politicians had frequently stated that giving more water to Syria and Iraq is impossible, in addition to being unacceptable. For instance, Taner Yildiz, the Turkish Energy Minister stated in 2009 that "[Türkiye] is aware of the need for water in [its] neighbors, but [it] does not have a lot of it in [its] reservoirs"[129] In this context, Syria and Iraq complained that Türkiye did not keep its promises regarding the accepted amount of water flow. Ahmet Davutoğlu, the Foreign Minister of Türkiye at the time, countered this argument by arguing that "Iraq's suffering in its lack of water is the same suffering Türkiye is going through. Turkish farmers have the same problems as Iraqi farmers."[130] Nevertheless, Syrian and Iraqi officials occasionally imply that Türkiye has enough rain and snow to fill its reservoirs and not to allow more water to flow to Syria and Iraq is rather a political choice.[131]

Some experts in the field even suggest that Türkiye's hard line position regarding the usage of its water sources by other countries has been strengthened through the "Neo-Ottomanist" politics of the AKP government and can even be compared to the energy hegemony of Russia over Europe with Türkiye trying to solidify its position as a "hydrological superpower."[132] Hussein Amery, a water specialist at the

Colorado School of Mines, says that Türkiye is a water superpower, because headwaters of the Tigris and Euphrates are in Turkish borders, which allows Türkiye to exert political power on the Arab world.[133] In contrast, however, other experts such as Gün Kut indicate that the real problem lies in the bad water management in the Arab World, as well as the rapidly rising population, which makes it difficult for Syria and Iraq to cope with food and water problems.[134] Moreover, war and domestic turmoil has been exacerbating the situation, further instigating migration out of these conflict zones with minimal human security and scarce natural resources.

Türkiye's ambitions to boost its economic growth with hydraulic based development projects increases the tensions between Türkiye and its Middle Eastern neighbors. The Guardian noted that numerous dam and hydropower schemes are being built on almost all of the main rivers in a drastic push to make Türkiye a world economic power by the centenary of the Republic in 2023.[135] Also, the AKP is particularly interested in reaching economic growth in poorer parts of Anatolia and thus to empower the "Anatolian tigers." The Turkish Prime Minister Erdoğan argued that rivers must no longer "run in vain" and 100% of its hydroelectric potential be harnessed over the next 12 years.[136] This statement has made environmentalists most worried. These developments have become a rather contentious issue at the heart of the energy-water nexus. Moreover, there is also some opposition and protests particularly at the local level.

While major water disputes regarding Türkiye have emerged in Eastern Anatolia, these problems affect Türkiye's relations with the EU as well. For instance, the EU noted that Türkiye uses its water resources as a weapon to increase its political power.[137] The EU Water Framework Directive requires member or candidate states to work with their co-riparians on water issues.[138] Additionally, Türkiye's river and dam policies raise an important question within the EU circles that Türkiye is not paying enough attention to human rights, cultural heritage, and the environment, which puts additional pressure on Türkiye's already lukewarm relations with the EU.[139] The lack of women's involvement in water management in Türkiye is another important concern. A Turkish environmental engineer, Arzu Özyol, states that women use water in households and daily activities much more frequently than men, and, therefore, should play an important role in water management. Despite their knowledge on water related issues, however, they are not given the incentive to take part in creating water policies.[140] Overall, Türkiye needs a serious assessment of its energy strategy in terms of its economic development aspirations, geopolitical priorities, and environmental concerns about water.

In the economy-environment-security triangle, its water resources are particularly important for Türkiye. The economic dimension has been characterized by Türkiye's comparative superiority of freshwater resources in the Middle East, and its aims to develop the Southeastern region of Türkiye and the development policy goals for the year 2023. Starting from the 1980s, Türkiye's Southeastern Anatolia Project has aimed at achieving local economic and social development, initially through farming with extensive irrigation systems and electricity generation, and later the scope has expanded to cover agriculture, industry, transportation, urban and rural infrastructure, health care, education, housing, and tourism. As previously indicated Türkiye's economic goals for 2023 gave the government a stimulus to expand its energy production, in which case rivers and their hydroelectric potential is viewed as a promising source. In response, environmentalists argue that the hydroelectric energy potential is low compared to the needs and potential environmental consequences are irreversible.[141]

In addition to development goals, the most important anchor point between water and environment is the climate change, which is expected to reduce precipitation in most regions of Türkiye. Thus, climate change will severely impact rain fed agriculture and therefore increase the agricultural and urban demand for water.[142] As indicated by the OECD, Türkiye has shown significant progress to fight against this issue through reforestation in recent years.[143] UN also certified Türkiye because of the successful reforestation and ranked Türkiye first in Europe and sixth in the world for the period between 2015 and 2020.[144] Concerning the efforts to fight climate change, access to wastewater treatment has also increased, although it is not still at an expected level.[145] Implication of climate change will be significant in the water sector.[146] Therefore, Türkiye needs improvements at the local level considering that municipal waste, which has exceeded 90%, is landfilled.[147] Apart from reforestation, Türkiye has been taking steps regarding water waste treatment, the Energy Efficiency Action Plan and renewable energy auctions. On the climate action front, Turkish Parliament's ratification of the Paris Agreement in 2021 has been a critical milestone. Subsequently, Türkiye announced a series of climate policies, from an unexpected 2053 carbon neutrality target, alongside a national Green Development Initiative, to the development of a National Green Finance Strategy by the end of 2023.

It is also important to stress water quality. The wetlands of Türkiye are on two major bird migration routes. The most significant of these wetlands are Göksu Delta, Kizilirmak Delta, Sultan Marshes, and Kus Cenneti. In fact, there is an environmental danger regarding these wetlands because of intentional draining to reclaim for agriculture and malaria control.[148] In areas where agrochemicals are extensively used, the hazardous effects of pesticides and fertilizers threaten the use of groundwater sources for drinking water. In the agricultural plains of Bornova (Izmir), the excessive use of agrochemicals resulted in significant groundwater pollution, with nitrate concentrations in the groundwater reaching the limit value of 45 mg/l.[149] Similarly, in the Nevşehir-Niğde provinces in Middle Anatolia, where 25% of the total potato growing area and 44% of total produce are located, groundwater resources and soils are seriously polluted with nitrate concentration. Various proportions of pesticide residues are encountered in drains, irrigation canals, small bays, some lakes, and in well water. Pesticide use in Türkiye is the highest in the Mediterranean region, particularly in the Çukurova region south of Adana. The Black Sea and Marmara Sea are also becoming polluted with agricultural pesticides. Some rivers and creeks such as the Ankara stream, the Ergene River and its branches in the Thrace region, Karasu creeks, a branch of the Sakarya River in Eskisehir, and the Simav stream in the Aegean region are all polluted by industrial, municipal, and agricultural wastewater.[150]

An alarming development, originating from the combined impact of extensive pollution and climate change, has been the "sea snot" disaster in the Marmara Sea. There have been regular mucilage problems in the Marmara Sea since 2007; however, the recent outbreak is the "biggest one yet."[151] During the summer of 2021, the Sea has been covered in sea snot which threatens marine life and the fishing industry in the region. Fishing declined 90% compared to January and as of June 19, 2021, more than 4,500 cubic meters of mucilage has been extracted.[152] The Minister of Environment expects the cleaning of mucilage from Marmara Sea within three years if the ongoing projects are successfully implemented,[153] with which Marmara Sea would become a protected area to avoid pollution and manage waste water coming from coastal cities and ships which caused the spread of the sea snot.[154] However, experts expect that mucilage will be present once every five or six years from now on and therefore urge the authorities to save the Black Sea and the Aegean.[155] In June 2021, 1,000 people and more than 70 ships participated in the cleaning operation at 77 different locations of the Marmara Sea.[156]

Agriculture emerges as a decisive factor for water usage. In fact, the semi-subsistence farms in Türkiye are characterized by low efficiency. The inefficiency in the agricultural sector is also reflected in the usage of high amounts of water that far exceed the OECD average.[157] Annually, approximately 44 billion m^3 of freshwater is used in Türkiye, while 74% is utilized for agriculture.[158] Yet, the global agricultural water usage is around 70% and 40% in many OECD countries.[159] Overall, Türkiye's economy-environment-security triangle clearly reflects the profile of Türkiye: a medium-size country prioritizing aspirations of geopolitical influence, and energy interests over longer-term environmental and sustainability concerns.

On the economic and environmental dimension of water issues, Azerbaijani authorities have started extensive collaboration with their Asian counterparts. For instance, the Asian Development Bank approved a 300-million-US-dollar loan to complete an urban water supply and sanitation project in Azerbaijan.[160] The project is designed for the improvement of water quality and sewerage services in Agdash, Goychay, Nakhchivan, and Beylagan towns, where poor sanitation caused severe public health and environmental problems. To address these problems, the Asian Development Bank had also assigned its first 30-million-dollar loan.[161] In December 2010, South Korean aid agency announced that it would lend a helping hand to the country plagued with drinking water shortages and water contamination with a monetary assistance of 80 billion won.[162] Despite these developments, the overall picture was still quite discouraging. Fazail Aghamaly, a deputy of Milli Majlis, claimed that the decline of water level of Lake Urmia "after two or three years" will mean that Azerbaijan itself will begin to drown in dust, with this tragedy hitting Nakhchivan first.[163]

Azerbaijan also suffers from a high level of water pollution. The Kura and Aras are highly polluted in Armenia and Georgia before they reach Azerbaijan. The Kura-Aras watershed has been characterized by intense industrial and agricultural pollution and even radioactive contamination.[164] Similarly, the environmental degradation in the Caspian Sea is one of the highest in the world that dates to the Soviet times. Experts from United Nations Environmental Programme point out that for Azerbaijan almost 30% of the Caspian Sea coastal area is exposed to contamination, more than half of the rivers are considered to be contaminated, and all the lakes in the low-lying parts of the country are being exposed to changes in thermal, biological, and chemical regimes.[165] Water shortages in Azerbaijan are expected to exceed 250,000 and 300,000 ha, which could then cause a decline in crop yields.[166] Pollution could even worsen this situation considering that water per capita could decrease by over 60%.[167] Particularly, problems in groundwater pollution and contamination are prevalent in the Absheron Peninsula, related mostly to oil production.[168] More importantly, 8.2 million people lack easy access to drinkable water in Azerbaijan which shows the poor water quality.[169] The Caspian Sea is also facing the problem of pollution and is in a "critical condition" considering that oil tankers annually drain more than 120,000 tons of pollutants, while sewage from cities in Iran, Russia, Azerbaijan, Kazakhstan, and Turkmenistan worsen the pollution.[170]

In the case of rivers and the Caspian Sea, there is a multitude of riparian countries, which implies that a solution is only possible through international political cooperation. Although countries in the region blame each other for the pollution of water sources, as in the case of Kura and Aras rivers between Armenia, Azerbaijan, and Georgia, some authorities—such as the officials responsible for the South Caucasus River Monitoring (SCRM) project—point out that focusing on improving access to fresh water may help states to develop contacts and practices that can solve other conflicts between nations.[171] For instance, Anna Gekht of Parliamentarians for Global Action (PGA) argued that by participating in forums organized by PGA, Aliyev and Kocharian have established a long working relationship and the forums have resulted in building a close personal relationship among the participating parliamentarians.[172]

The environmental problems have a great impact on the freshwater resources of Azerbaijan. Erosion dramatically affects the irrigation infrastructure, a great portion of which is in dire need of renovation. In addition, wastewater treatment facilities continue to pose a problem. Mixing industrial and municipal wastewater prior to treatment is common and wastewater is usually directly pumped into the sea or storage lakes. In 2015, Azerbaijan completed construction of the world's largest surface water filtration plant in Baku, under plans developed jointly by HidroLutus, a water infrastructure building company and Dow Water Systems.[173] The plant uses nanotechnology filters evaluated by key research centers in Türkiye and the United States.[174] The Asian Development Bank reportedly is set to invest in substantial water treatment modernization projects in Tbilisi soon as well.

The energy-water nexus is also intricately intertwined with food security. Türkiye mainly prioritizes "food and nutrition security and food safety; sustainable forest and natural resources management, including fisheries; and institutional capacity enhancement of public and private sectors."[175] Türkiye has partnered with the United Nations Food and Agriculture Organization (FAO), for food security and sustainable management of natural resources.[176] As part of the FAO's work, Türkiye spent more than USD 75 million for making agriculture, forestry, and fisheries more productive and sustainable, enabling inclusive and efficient agricultural and food systems.[177] Additionally, Türkiye mainly funded projects to support vulnerable populations in Central Asia.

In comparison, the South Caucasus has more problems regarding food security. Armenia is facing challenges in nutrition,[178] because of the landlock status of the country, weather shocks, and the climate change which adversely affect the economy and the nutrition.[179] 15% of the population in Armenia is food insecure and 23.5% live in poverty.[180] The COVID-19 pandemic has also affected food supply and demand, and decreased purchasing power and the capacity to produce and distribute food.[181] In Georgia, land resources are limited since only 15% of the land is cultivated.[182] Hence, a project was initiated by the United Nations Environment Programme (UNEP) for sustainable land management in Georgia through improving soil productivity and food security with crop rotation and inter-cropping in Gori, Kareli, Kvareli, and Sagarejo municipalities.[183] Azerbaijan, on the other hand, ranked 56th out of 113 in the global food security index in 2018.[184] The natural soil and climatic conditions in Azerbaijan are favorable for organic farming and thereby the country works on this to supply the domestic and export markets.[185] Although Azerbaijan has experienced farmers as well as scientific and educational institutions providing trainings and expertise, the agricultural sector constitutes 6% of the GDP.[186] Lack of modernization and low productivity because of limited rainfall, arable land and fragmented landholdings, and soil erosion increases challenges regarding food security. The Russian invasion of Ukraine has also aggravated problems regarding food security in this region and beyond.

Conclusion

In this chapter, we explored Türkiye's interactions with the Greater Middle East in general, and with the South Caucasus in particular, regarding the energy-water-food nexus and examined how these complex webs of relations shape our broader understanding of security. We posited that while there has been some identity based and ideological factors reflecting the pan-Turkist and neo-Ottomanist trends in Turkish foreign policy, energy interests, and geopolitical concerns have emerged as prevalent factors shaping Türkiye's approach towards this region. The environmental and sustainability dimension, however, have often been neglected by all regional actors. Moreover, there are serious challenges concerning the governance of transboundary natural resources in a turbulent region torn by territorial disputes and international conflict aggravating the tensions.

The South Caucasus have suffered from the heritage of Soviet indifference to environmental degradation and impact. Large-scale oil and gas extraction operations in Azerbaijan, throughout the 20th century, have left their mark in the heavily polluted ground and surface waters. Large-scale hydroelectric projects in the Republic of Georgia remain as challenges for repair and operation particularly in areas of tension. Substandard sewage treatment and pollution control systems from the Soviet period were neglected through much of the early transition years in both countries, given other political and economic priorities and financial constraints.

In the case of the Kura-Aras problem, Georgia, Armenia, and Azerbaijan have refused to cooperate on the management of the Kura-Aras and often blame each other for dumping pollutants into its waters. This region is especially significant for the Caucasus since a substantial part of the population of South Caucasia lives in the Kura and Aras River basin and most of the agricultural and industrial enterprises are located on their banks. The political tensions run so deep that when NATO announced a water project in the basin, the three governments took out the word "cooperation" from its title.[187] Azerbaijani scientists had to get approval from the Ministry of Foreign Affairs to meet with their counterparts from neighboring Armenia. This case also presents a good example of Putnam's two-level game conception as discussed in the theory section of Chapter 1, where the domestic determinants could also turn out to be a critical obstacle towards cooperation and affect international bargaining.

Many experts view that water is not likely to trigger large-scale violence; however, as the relations between Armenia, Georgia, and Azerbaijan are already very fragile, this problem can exacerbate existing tensions with far-reaching effects for the region.[188] Matanat Avazova from the Azerbaijani Ministry of Ecology claimed that the Kura and Aras were being contaminated by Armenia, Georgia, and Iran.[189] Georgian Green Party Head Georgi Gachechiladze argued that "an artificially created water deficit might cause tension between the states of [Georgia, Azerbaijan and Türkiye.]"[190] There is, however, also increasing cooperation among some of the regional actors. There are intensifying energy ties between Türkiye and Azerbaijan, enhancing energy security for these countries and Europe. As the Russian invasion of Ukraine has once again demonstrated the significance of diversification for European energy security, energy supply routes going through the Caucasus and Türkiye will maintain their significance.

As discussed above, many environmental remediation projects regarding water resources are also in place in different parts of this region, thanks to funding from a range of international donors, and new commitment from national government leadership. One can see the impact of these initiatives, for instance in the new attention to sewage treatment in Baku and Tbilisi, as well as in the modification of hydroelectric power plant proposals to scale down the environmental and social impact.

In conclusion, expanding economic growth and addressing energy security challenges, while addressing climate change in the face of dwindling water resources remains a daunting task both at the domestic and international level. The environmental degradation in the region is striking and needs more comprehensive measures and uneasy transnational cooperation in a turbulent setting. There is a dire need for better governance based on transparency, respect for and compliance with legal and regulatory frameworks as well as international agreements, institutionalized mechanisms for coordination and data sharing, and the inclusion of all stakeholders. These measures also have significant human security implications at the local level. The imperatives remain powerful, and the degree of commitment to remediation from national leaders remains uncertain, yet crucial for future environmental sustainability in the region.

Notes

1 For a more extensive discussion on these concepts, see: Jacob M. Landau, *Pan-Turkism: From Irredentism to Cooperation* (Bloomington: Indiana University Press, 1995); and M. Hakan Yavuz, *Nostalgia for the Empire: The Politics of Neo-Ottomanism* (Oxford: Oxford University Press, 2020).

2 Hoffman in Ebel and Menon, 2000, pp. 55–70; See also: Feshbach and Friendly, 1992 for a broader discussion of Soviet "Ecocide" and Kaplan, 1996, pp. 163–172, and Goltz, 1998, for interesting descriptions of the hard life in Baku in the early 1990s.

3 Central Intelligence Agency, "Explore All Countries Azerbaijan," June 8, 2021. https://www.cia.gov/the-worldfactbook/countries/azerbaijan/#energy

4 Hoffman in Ebel and Menon, 2000, p. 71.

5 LeVine, 2007.

6 Reuters, "Georgian Tycoon Starts Cabinet Talks, Prioritizes U.S. Ties," October 3, 2012. https://www.reuters.com/article/georgia-election-idUSL6E8L335820121003

7 Georgia Ministry of Agriculture, "Strategy for Agricultural Development in Georgia 2015–2020," 2015. http://extwprlegs1.fao.org/docs/pdf/geo145265.pdf

8 See: University of Gothenberg. School of Business, Economics and Law, "Georgia Environmental and Climate Change Policy Brief," January 22, 2009. Retrieved July 1, 2021, from http://sidaenvironmenthelpdesk.se/wordpress3/wp-content/uploads/2013/04/Georgia-EnvCC-Policy-Brief-Draft 090130.pdf

9 IW: Learn, "Georgia and Azerbaijan Share Rivers and Plans for a Water Rich Future," April 19, 2017. https://news.iwlearn.net/georgia-and-azerbaijan-share-rivers-and-plans-for-a-water-rich-future

10 See Danielyan, 2013 for an interesting interpretation of the Armenian decision.

11 Katerina Dalacoura, "Turkish Foreign Policy in the Middle East: Power Projection and Post-Ideological Politics," *International Affairs* 97, no. 4 (2021): pp. 1125–1142. doi:10.1093/ia/iiab082

12 For a comprehensive account of multiple dimensions of Azerbaijan-Türkiye Relations, see Murad Ismayilov and Norman A. Graham (eds.), *Turkish-Azerbaijani Relations: One Nation-Two States?* (New York: Routledge, 2016).

13 BP, "BP Statistical Review of World Energy," June 2020, 2020. https://www.bp.com/content/dam/bp/business-sites/en/global/corporate/pdfs/energy-economics/statistical-review/bp-stats-review-2020-full-report.pdf

14 BP, "BP Statistical Review of World Energy," June 2022, 2022. https://www.bp.com/content/dam/bp/business-sites/en/global/corporate/pdfs/energy-economics/statistical-review/bp-stats-review-2022-full-report.pdf

15 Doğalgaz Piyasası Daire Başkanlığı, "Doğalgaz Piyasası 2020 Yılı Sektör Raporu," (Ankara, 2021), s. 10.
16 International Energy Agency. https://www.iea.org/news/how-europe-can-cut-natural-gas-imports-from-russiasignificantly-within-a-year. (Latest Access March 24, 2022).
17 Marco Siddi, "The Scramble for Energy Supplies to Southeastern Europe: The EU's Southern Gas Corridor, Russia's Pipelines and Türkiye's Role," in *Türkiye as an Energy Hub?: Contributions on Türkiye's Role in EU Energy Supply*, (2017).
18 For a detailed discussion on the asymmetric dependence between Russia and Türkiye, see Ziya Öniş and Şuhnaz, "Türkiye and Russia in a Shifting Global Order: Co-Operation, Conflict and Asymmetric Interdependence in a Turbulent Region," *Third World Quarterly* 37, no. 1 (January 2016): pp. 71–95.
19 Dünya, "Cari açığın en önemli sebebi enerji ithalatı," January 15, 2017. https://www.dunya.com/sektorler/enerji/cari-acigin-en-onemli-sebebi-enerji-ithalati-haberi-345819
20 Ebru Şengül, "Türkiye's Energy Imports Fall by 28.2% in 2016." *Anadolu Agency*, February 1, 2017. http://aa.com.tr/en/economy/Türkiyes-energy-imports-fall-by-282-in-2016/739208
21 International Energy Agency, Türkiye 2021 Energy Policy Review. https://iea.blob.core.windows.net/assets/cc499a7b-b72a-466c-88de-d792a9daff44/Türkiye_2021_Energy_Policy_Review.pdf
22 Haberturk, "Azerbaycan'dan doğalgaz misillemesi," April 25, 2009. http://ekonomi.haberturk.com/makroekonomi/haber/142861-azerbaycandan-dogalgaz-misillemesi
23 IHS Global Insight, "Azerbaijan, Türkiye Agree on Gas Price, Supply, and Transit Terms," June 8, 2010. http://www.ihs.com/products/global-insight/industry-economicreport.aspx?ID=106594168&pu=1&rd=globalinsight_com
24 Meltem Muftuler-Bac, "The Future of Energy Security for Europe: Türkiye's Role as an Energy Corridor," *Middle Eastern Studies* 47, no. 2 (2011): pp. 361–378.
25 For a comprehensive assessment of challenges concerning both projects, see: Pavel K. Baev and Indra Overland, "The South Stream versus Nabucco Pipeline Race: Geopolitical and Economic (Ir)Rationales and Political Stakes in Mega-Projects," *International Affairs* 86, no. 5 (2010): pp. 1075–1090.
26 Azer News, "TANAP Project to Be Completed Earlier than Planned, Minister Says," October 6, 2017. https://www.azernews.az/oil_and_gas/120096.html; and Turkish Foreign Ministry. 2019. http://www.mfa.gov.tr/Türkiyes-energy-strategy.en.mfa
27 Jonathan Levack, "South Stream and the Trans-Anatolian pipeline: The End of Nabucco?" *SETimes*, January 11, 2012. http://www.setimes.com/cocoon/setimes/xhtml/en_GB/features/setimes/features/2012/01/11/feature-04
28 Şuhnaz Yılmaz and Duygu Sever-Mehmetoğlu, "Linking Foreign Policy and Energy Security: An Asset or a Liability for Türkiye," *Uluslararası İlişkiler* 13, no. 52 (2016): p. 119.
29 Milliyet, "Nabucco Projesi Tarih Oldu," November 18, 2013. www.milliyet.com.tr/nabucco-projesi-tariholdu/ekonomi/detay/1793995/default.htm
30 Petroleum Pipeline Corporation (BOTAŞ), 2022. https://www.botas.gov.tr/pages/natural-gas-and-crude-oil-pipeline-map/416
31 Republic of Türkiye Energy Market Regulatory, "Enerji Piyasası Düzenleme Kurumu, Doğalgaz Piyasası Sektör Raporu" (Ankara, 2021), p. 10.
32 Milliyet, "Rusya'ya 'yılbaşı hediyesi' Güney Akım Boru Hattı izni," *Milliyet*, December 2011. http://www.milliyet.com.tr/rusya-ya-yilbasi-hediyesi-guney-akim-boru-hatti-izni-ekonomi-1481725/
33 NTV, "Rusya Akkuyu nükleer santrali için tarih verdi," July 10, 2017. https://www.ntv.com.tr/ekonomi/rusyaakkuyu-nukleer-santrali-icin-tarih-verdi,7YKGOMoiUEijgSFG3sUxEg
34 Emre Erşen and Mitat Çelikpala, "Türkiye and the Changing Energy Geopolitics of Eurasia," *Energy Policy* 128 (2019): pp. 584–592.

35 Hürriyet Daily News, "Asian Infrastructure Investment Bank Approves $600 Mln Loan for Azerbaijan-Türkiye Gas Pipeline," December 22, 2016. www.hurriyetdailynews.com/asian-infrastructure-investment-bank-approves-600-mln-loan-for-azerbaijan-Türkiye-gas-pipeline-107613
36 Daily Sabah, "Türkiye, China Partner Up for 1.2-GW Solar Plant in Gebze in 2017," July 14, 2017. https://www.dailysabah.com/energy/2017/07/14/Türkiye-china-partner-up-for-12-gw-solar-plant-in-gebze
37 Burak Bir, "Food Insecurity Türkiye's Top Climate Change Risk." *Anadolu Agency*, May 6, 2021. https://www.aa.com.tr/en/environment/food-insecurity-Türkiyes-top-climate-change-risk/2230992
38 Republic of Türkiye Ministry of Foreign Affairs, "Türkiye's Policy on Water Issues." https://www.mfa.gov.tr/Türkiye_s-policy-on-water-issues.en.mfa
39 Loretta Martikian, "EBRD and EU Join Forces to Help Georgia's Private Sector Seize New Opportunities." *EBRD*, May 27, 2016. http://www.ebrd.com/cs/Satellite?c=Content&cid=1395250835835&d=Mobile&pagename=EBRD%2FContent%2FContentLayout
40 The Asian Development Bank, "Member Fact Sheet: Georgia," 2020. https://www.adb.org/sites/default/files/publication/27765/geo-2020.pdf
41 Ibid.
42 Georgian State Electrosystem, "Construction of Substation Khorga 220/110kV Has Been Completed," June 3, 2016. http://www.gse.com.ge/communication/news/2016/construction-of-substation-khorga-220/110kv-hasbeen-completed
43 See ADB reports, 44183-013, 47919-01.
44 See World Bank Reports: P054886, P072394, P064094.
45 World Bank, "World Bank in Georgia: 25 Years of Partnership," September 15, 2017. http://www.worldbank.org/en/news/feature/2017/08/24/world-bank-georgia-25-years-of-partnership
46 The Asian Development Bank, "Country Report: Assessment of Power Sector Reports in Georgia," 2015. https://www.adb.org/sites/default/files/institutional-document/161067/georgia-psr-report.pdf
47 Ibid.
48 U.S. Agency for International Development (USAID), "Georgia: Energy," 2017. https://www.usaid.gov/georgia/energy
49 Caucasus Business Week, "Khudoni Hpp – Project of the Century for Georgia," 2015. http://cbw.ge/business/khudoni-hpp-project-of-the-century-for-georgia/
50 Civil Georgia, "EIB to Allocate EUR 250 Million to Georgia," July 13, 2017. *EIB to Allocate*. http://www.civil.ge/eng/article.php?id=30264
51 See Popescu, 2007, p. 30.
52 Azeri News Trend, October 5, 2012. http://en.trend.az/capital/energy/2073769.html
53 "The Messenger Online," 15 January 2009. http://www.messenger.com.ge/issues/1774_january_15_2009/1774_enguri.html
54 Civil Georgia, "Enguri HPP Resumes Operation," March 7, 2017. http://www.civil.ge/eng/article.php?id=29911
55 Caucasus Business Week, "Khudoni Hpp – Project of the Century for Georgia," 2015. http://cbw.ge/business/khudoni-hpp-project-of-the-century-for-georgia/
56 Trend News Agency, October 5, 2012.
57 See Natia Ejoshvili, "Dam Will Reshape Remote Georgian Mountain Region." Transitions Online, June 29, 2011. http://www.tol.org/client/article/22514.html; and Azeri News Trend, October 5, 2012.
58 Georgian Young Lawyers Association, "NGO Statement on the Khodoni Hydroelectric Power Plant Project," October 17, 2013. Retrieved July 1, 2021 from http://gyla.ge/eng/news?info=1774
59 Power Insider, April 5, 2013.

60 Business Media Georgia, "Government Continues Negotiations with Transelectrica to Terminate Khudoni HPP Agreement," December 7, 2020. https://bm.ge/en/article/government-continues-negotiations-with-transelectricato-terminate-khudoni-hpp-agreement/70482/
61 Trend news Agency, "Georgia Considering Construction of Khudoni HPP," June 28, 2021. https://en.trend.az/business/energy/3446823.html
62 Ajaratv.ge, May 14, 2012.
63 See http://www.adjaratv.ge/en/index.php?page=shownews&id=23289
64 United Nations, "United Nations Sustainable Development Goals: Ensure Access to Affordable, Reliable, Sustainable and Modern Energy." www.un.org/sustainable development/energy/
65 International Energy Agency, "Türkiye 2021 Energy Policy Review," 2021. p. 95. https://iea.blob.core.windows.net/assets/cc499a7b-b72a-466c-88ded792a9daff44/Türkiye_2021_Energy_Policy_Review.pdf
66 General Directorate of Renewable Energy, "Türkiye," 2016. http://www.eie.gov.tr/document/2016_f_r.pdf
67 Türkiye Kömür İşletmeleri Kurumu. "Kömür Ve Getirdiği Çözümler." http://www.tki.gov.tr/dosyalar/temiz_komur.pdf.
68 International Clean Coal. "Clean Coal Summit 2017 in Türkiye." 2017. http://komur-der.org/tr/files/uluslararasirapor.pdf, November 11, 2017.
69 Republic of Türkiye Ministry of Energy and Natural Resources, 2022. http://www.enerji.gov.tr/enUS/Pages/International
70 International Energy Agency, "Key World Energy Statistics 2017." (Paris, 2017), pp. 59–69.
71 Energy Information Administration (EIA), Country Analysis Executive Summary: Azerbaijan, 2021. https://www.eia.gov/international/analysis/country/AZE.
72 Energy Charter Secretariat, "Azerbaijan: In-depth Review of the Investment Climate and Market Structure in the Energy Sector." (Brussels, 2011), pp. 79–80.
73 Jamil Malikov and Parvin Mammadzadeh, "Sustainable Development in Energy Sector of Azerbaijan." *UNECE*, October 28, 2016. www.unece.org/fileadmin/DAM/energy/se/pp/eneff/7th_IFESD_Baku_Oct.2016/8_path/Malikov_Azerbaijan.p df.
74 The Ministry of Energy of the Republic of Azerbaijan, "The Use of Renewable Energy Resources in Azerbaijan," April 7, 2020. https://minenergy.gov.az/en/alternativ-ve-berpa-olunan-enerji/azerbaycanda-berpa-olunan-enerjimenbelerinden-istifade
75 World Energy Council, Energy Efficiency Policies Around the World: Review and Evaluation 2008 (London: 2007), pp. 8–9.
76 Chris Green, "Energy Intensity: Complexities, Implications, Controversies" (Energy Institute, University of Wisconsin-Madison, March 2011). http://www.energy.wisc.edu/wp-content/uploads/2011/02/Energy-IntensityRevFinal.pdf.
77 Erdal Çalıkoğlu, "Energy Efficiency Policies and Programs in Türkiye" (Elektrik İşleri Etüt İdaresi Genel Müdürlüğü, 2010). http://siteresources.worldbank.org/EXTENERGY2/Resources/41141991276110591210/Türkiye.pdf
78 International Energy Agency, *Key World Energy Statistics 2017*, pp. 59–69. https://www.iea.org/publications/freepublications/publication/KeyWorld2017.pdf
79 Erdal Çalıkoğlu, "Türkiye'de Yenilenebilir Enerji Ve Enerji Verimliliği." Yeşil Enerji'de Danimarka ile Fırsatlar Konferansı, İstanbul, March 2, 2012.
80 Yusuf Yazar, Türkiye'nin Enerji Durumu ve Geleceği, 2010.
81 YEGM, "Enerji Verimliliği Strateji Belgesi 2012–2023." *Resmi Gazete*, February 2012. https://www.resmigazete.gov.tr/eskiler/2012/02/20120225-7.htm
82 International Energy Agency, "Key World Energy Statistics 2017." https://www.iea.org/publications/freepublications/publication/KeyWorld2017.pdf
83 Ibid.

84 European Bank of Reconstruction and Development, "Azerbaijan Country Profile." http://www.ebrd.com/downloads/legal/irc/countries/azerbaijan.pdf

85 European Bank for Reconstruction and Development, "Commercial Laws of Azerbaijan – An Assessment by EBRD." June 2014. www.ebrd.com/.../legal-reform/read-the-ebrds-azerbaijan-country-law-assessment.pdf

86 Rafig Verdiyev, "Water Resources Development and Management in Azerbaijan." *United Nations Economic and Social Commission for Asia and the Pacific*, 2008. http://www.unescap.org/

87 Aquastat, Azerbaijan, 2008. http://www.fao.org/nr/water/aquastat/countries_regions/AZE/

88 Ibid.

89 United Nations, "Implementing Water." 2015. http://www.un.org/waterforlifedecade/waterandsustainabledevelopment2015/images/implementing_water _eng.pdf

90 Rafig Verdiyev, "Water Resources Development and Management in Azerbaijan." *United Nations Economic and Social Commission for Asia and the Pacific*, 2008. http://www.unescap.org/

91 United Nations, "UN World Water Development Report." March 16, 2009. https://www.unwater.org/publications/water-changing-world/

92 World Health Organization, "UN-Water Global Analysis and Assessment of Sanitation and Drinking-Water." 2015. http://www.who.int/water_sanitation_health/monitoring/investments/azerbaijan-20-oct-16.pdf?ua=1

93 United Nations Economic Commission for Europe, "Second Environmental Performance Review." 2004. https://unece.org/DAM/env/epr/epr_studies/azerbaijan%20II.pdf

94 Ibid.

95 Index Mundi, "Türkiye Area." November 27, 2020. https://www.indexmundi.com/Türkiye/area.html); and Indexmundi, "Greece Area." November 27, 2020. https://www.indexmundi.com/greece/area.html

96 Knoema, "Dependency ratio(%)." https://knoema.com/atlas/ranks/Dependency-ratio

97 Samira Ahmedbeyli, "Azeris Accuse Karabakh of Denying Them Water." September 24, 2010. https://iwpr.net/global-voices/azeris-accuse-karabakh-denying-them-water

98 Karine Ohanyan, "Water Complicates Karabakh Peace Talks." September 17, 2010. http://iwpr.net/reportnews/water-complicates-karabakh-peace-talks/

99 Ibid.

100 Hürriyet Daily News, "Life Returns to Normal in Terter after 30 Years of Occupation." https://www.hurriyetdailynews.com/life-returns-to-normal-in-terter-after-30-years-of-occupation-162693

101 TUSIAD, "Water Report 2008." http://www.tusiad.org:7979/FileArchive/su_yonetimi.pdf/

102 Water Fanack, "Surface Water and Groundwater Resources in Türkiye." May 4, 2021. https://water.fanack.com/Türkiye/Türkiye-surface-water-groundwater/

103 Ibid.

104 Aquastat. "Türkiye." 2008. http://www.fao.org/nr/water/aquastat/countries/Türkiye/index.stm/); and The Republic of Türkiye-Ministry of Foreign Affairs, "International Treaties." http://ua.mfa.gov.tr

105 The Republic of Türkiye-Ministry of Foreign Affairs, "Water: A Source of Conflict of Cooperation in the Middle East?" https://www.mfa.gov.tr/data/DISPOLITIKA/WaterASourceofConflictofCoopintheMiddleEast.pdf

106 Aquastat, "Türkiye." 2008. http://www.fao.org/nr/water/aquastat/countries/Türkiye/index.stm/

107 World Bank, "People Using at least Basic Sanitation Services, Urban (% of Urban Population) – Türkiye." December 10, 2017. https://data.worldbank.org/indicator/SH.STA.BASS.UR.ZS?locations=TR

108 Index Mundi, "Türkiye Sanitation Facility Access." https://www.indexmundi.com/Türkiye/sanitation_facility_access.html

109 DSI, 2006.

110 World Bank, "Türkiye Irrigation Modernization Project (P158418)." November 2, 2018. https://documents1.worldbank.org/curated/pt/990411541777029255/pdf/Project-Information-Document-Integrated-Safeguards-Data-Sheet-Türkiye-Irrigation-Modernization-P158418.pdf

111 B. Sönmez. *Türkiye'de Çorak Islahi Arastirmalari ve Tuzlu Topraklarin Yönetimi.* Sulanan Alanlarda Tuzluluk Yönetimi Sempozyumu Bildiriler Kitabi. 20–21 Mayis. Ankara. 2004.

112 Water Fanack, "Water Infrastructure in Türkiye." November 21, 2016. https://water.fanack.com/Türkiye/waterinfrastructure-in-Türkiye/, January 12, 2018

113 PBS Global Connections, "The Middle East." October 2002. https://www.pbs.org/wgbh/globalconnections/mideast/questions/resource/index.html

114 Gal Luft, "Dependence on Middle East Energy and Its Impact on Global Security" Energy and Environmental Challenges to Security, pp. 197–210. https://link.springer.com/chapter/10.1007%2F978-1-4020-9453-8_13

115 Paul Cochrane, "Renewable Energy: What Does It Mean for Oil-Dominated Middle East?." *The Middle East Eye*, October 30, 2020. https://www.middleeasteye.net/news/renewable-energy-oil-middle-east

116 FAO, "Solutions to Water Challenges in the Middle East and North Africa region," August 30, 2021. http://www.fao.org/fao-stories/article/en/c/1150870/

117 Ibid.

118 Anders Jagerskog, "Water, Food, and Energy in the Arab World: A Collective Challenge." *World Bank*, November 20, 2018. https://blogs.worldbank.org/arabvoices/water-food-and-energy-arab-world-collective-challenge

119 Hakan Istanbulluoglu, "Water Policies of Türkiye." 2011. http://www.scopemed.org/?mno=4135/

120 Water Politics, *The Ring Tide of Mideast Water Tension: Neo-Ottomanism& Türkiye As a Hydrological Superpower*, January 9, 2012. http://www.waterpolitics.com/2010/01/09/the-ring-tide-of-mideast-watertension-neo-ottomanism-Türkiye-as-a-hydrological-superpower/

121 A. Akanda, S. Freeman, and M. Placht, *The Tigres-Euphrates River Basin: Mediating a Path Towards Regional Water Stability*, 2007.

122 I. Kaya, *The Euphrates-Tigris Basin: An Overview and Opportunities for Cooperation under International Law*, 1998.

123 A. Akanda, S, Freeman, and M. Placht, *The Tigres-Euphrates River Basin: Mediating a Path Towards Regional Water Stability*, 2007.

124 Food and Agricultural Organization, 2008. http://www.fao.org/nr/water/aquastat/countries_regions/TUR/

125 Gaafar Roba, "The Environmental Impact of Syria's Conflict: A Preliminary Survey of Issues." *Arab Reform Initiative*, April 7 2021. https://www.arab-reform.net/publication/the-environmental-impact-of-syrias-conflict-apreliminary-survey-of-issues/

126 Ibid.

127 Sirwan Kajjo and Ezel Sahinkaya, "Water Drop in Euphrates River Increases Tensions Between Syrian Kurds, Türkiye." Voice of America, May 6, 2021. https://www.voanews.com/extremism-watch/water-drop-euphratesriver-increases-tensions-between-syrian-kurds-Türkiye

128 Ibid.

129 Water Politics, "Türkiye Unable to Give more Water to Iraq and Syria." September 8, 2009. http://www.waterpolitics.com/2009/09/08/Türkiye-unable-to-give-more-water-to-iraq-and-syria/

130 Water Politics, "Iraq, Türkiye, and Water: More Than Just Surface Tension." August 13, 2009. http://www.waterpolitics.com/2009/08/13/iraq-Türkiye-and-water-more-than-just-surface-tension/

131 Sputnik News, "Thirst Wars: Türkiye Applies Political Pressure by Cutting Off Water to Syria." March 8, 2017. https://sputniknews.com/middleeast/201703081051370769-Türkiye-syria-water-euphrates/

132 Water Politics, "The Ring Tide of Mideast Water Tension: Neo-Ottomanism & Türkiye as a Hydrological Superpower." January 9, 2012. http://www.waterpolitics.com/2010/01/09/the-ring-tide-of-mideast-watertension-neo-ottomanism-Türkiye-as-a-hydrological-superpower/
133 Ibid.
134 Ibid.
135 Water Politics, "Türkiye's Great Leap Forward Leads to Water Tension." June 2, 2011. http://www.waterpolitics.com/2011/06/02/Türkiyes-great-leap-forward-plans-lead-to-water-tension/
136 Ibid.
137 Water Politics, "Türkiye's Contemplated European Union "Freeze" – All About Water?." February 3, 2012. http://www.waterpolitics.com/2012/02/03/Türkiye%E2%80%99s-contemplated-european-union%E2%80%9Cfreeze%E2%80%9D-all-about-water/
138 Ibid.
139 Conference on Türkiye's River and Dam Policy, November 18, 2010. http://m-h-s.org/ilisu/upload/PDF/2010/Invitation_Dam_Conference_Nov_2010.pdf/
140 İzgi Güngör. "Women Absent in Water Politics." *Hurriyet Daily News*, January 26, 2009. http://www.hurriyet.com.tr/english/domestic/10852777.asp/
141 Water Politics, "Türkiye's Great Leap Forward Leads to Water Tension." June 2, 2011. http://www.waterpolitics.com/2011/06/02/Türkiyes-great-leap-forward-plans-lead-to-water-tension/
142 Sector Note. "World Bank Türkiye – National Basin Management Strategy." December 11, 2010. www.gonder.org.tr/wp-content/uploads/2015/01/The-World-Bank-Sector_Note-on-WatershedManagement.pdf.
143 Ibid.
144 Hürriyet Daily News, "UN Recognizes Türkiye's Success in Forestation Efforts." February 25, 2021. https://www.hurriyetdailynews.com/un-recognizes-Türkiyes-success-in-forestation-efforts-162676
145 OECD, "OECD Environmental Performance Reviews: Türkiye 2019." https://www.oecd.org/Türkiye/oecdenvironmental-performance-reviews-Türkiye-2019-9789264309753-en.htm
146 OECD, "Preface." https://www.oecd-ilibrary.org/sites/9789264309753en/index.html?itemId=/content/publication/9789264309753-en
147 Ibid.
148 N. Harmancioglu, N. Alpaslan, and E. Boelee, *Irrigation, Health and Environment: A Review of Literature from Türkiye* (Colombo, Sri Lanka: International Water Management Institute (IWMI working paper 6), 2001), p. 21.
149 Ibid.
150 Aquastat, "Türkiye." 2008. http://www.fao.org/nr/water/aquastat/countries/Türkiye/index.stm/
151 The Economist, "Sea Snot" Clogs Up the Marmara Sea." June 26, 2021. https://www.economist.com/europe/2021/06/26/sea-snot-clogs-up-the-marmara-sea
152 Bilgiç Taylan, "Türkiye Races to Stop 'Sea Snot' Spreading to Neighboring Waters." Bloomberg Green. June 21, 2021. https://www.bloomberg.com/news/articles/2021-06-21/Türkiye-races-to-stop-sea-snot-spreading-toneighboring-waters
153 Voice of America, "Türkiye Says It Will Defeat 'Sea Snot' Outbreak in Marmara Sea." June 6, 2021. https://www.voanews.com/europe/Türkiye-says-it-will-defeat-sea-snot-outbreak-marmara-sea
154 Reuters, "Türkiye Says It Will Defeat 'Sea Snot' Outbreak in Marmara Sea." June 7, 2021. https://www.reuters.com/world/middle-east/Türkiye-says-it-will-defeat-sea-snot-outbreak-marmara-sea-2021-06-06/
155 Bilgiç Taylan, "Türkiye Races to Stop 'Sea Snot' Spreading to Neighboring Waters." *Bloomberg Green*, June 21, 2021. https://www.bloomberg.com/news/articles/2021-06-21/Türkiye-races-to-stop-sea-snot-spreading-toneighboring-waters

156 Climate Home News, "An Outbreak of 'Sea Snot' Is Killing Türkiye's Fishing Industry." June 18, 2021. https://www.climatechangenews.com/2021/06/18/outbreak-sea-snot-killing-Türkiyes-fishing-industry/
157 Kenan Demirkol, "Aykırı Sorular." CNN Turk, August 15, 2012. http://tv.cnnturk.com/video/2012/08/16/basliksiz/index.html
158 World Bank, "Republic of Türkiye Natural Capital Accounting Valuing Water Resources in Türkiye: A Methodological Overview and Case Study." September 2016. https://documents1.worldbank.org/curated/en/600681476343083047/pdf/AUS10650-REVISED-PUBLIC-Türkiye-NCA-Water-Valuation-Report-FINAL-CLEAN.pdf
159 OECD, "Managing Water Sustainably Is Key to the Future of Food and Agriculture." https://www.oecd.org/agriculture/topics/water-and-agriculture/
160 Asian Development Bank, "ADB Approves 300 million USD Loan for Azerbaijan's Water Supply Project. June 7, 2012. http://www.thefinancialexpress-bd.com/more.php?news_id=132268&date=2012-06-08/
161 Ibid.
162 Kang Hyun-kyung, "Green ODA to Ease Water Disputes in Azerbaijan." January 24, 2011. http://www.koreatimes.co.kr/www/news/nation/2011/01/116_80217.html
163 Paul Goble, "Azerbaijan Diplomatic Academy. Water as a Foreign Policy Challenge for Azerbaijan." October 1, 2011. http://biweekly.ada.edu.az/vol_4_no_19/Water_as_a_foreign_policy_challenge_for_Azerbaijan.htm
164 Oliver Bullough, "Armenian-Azerbaijani Water Politics." October 21, 2010. http://www.globalpolicy.org/component/content/article/198-natural-resources/49597-armenian-azerbaijaniwater-politics.html
165 Aquastat, "Azerbaijan." 2008. http://www.fao.org/nr/water/aquastat/countries_regions/azerbaijan/index.stm/
166 World Bank, "Climate Change and Agriculture Country Note." June 2012. p. 13. https://documents1.worldbank.org/curated/en/575751468014471558/pdf/954070WP0Azerb00Box391416B00P UBLIC0.pdf
167 Ibid.
168 Asian Development Bank, "Country Partnership Strategy: Azerbaijan, 2014–2018 – Azerbaijan: Country Environment Note." pp. 5, 8. https://www.adb.org/sites/default/files/linked-documents/cps-aze-2014-2018-sd04.pdf
169 The Borgen Project, "Water Quality in Azerbaijan." https://borgenproject.org/water-quality-in-azerbaijan/
170 The Guardian, "Pollution and Overfishing Plague the Caspian Sea." March 11, 2015. https://www.theguardian.com/world/iran-blog/2015/mar/11/iran-caspian-sea-pollution-overfishing
171 Oliver Bullough, "Armenian-Azerbaijani Water Politics." October 21, 2010. http://www.globalpolicy.org/component/content/article/198-natural-resources/49597-armenian-azerbaijaniwater-politics.html
172 Anna Gekht, "Sustainable Development & Population Programme", Parliamentarians for Global Action (PGA), OSCE/USAID Regional Workshop, May 16, 2003. http://www.worldwatercouncil.org/fileadmin/wwc/Library/Publications_and_reports/Proceedings_Water_Politi cs/proceedings_waterpol_pp.197-224.pdf
173 Tega, "Bakü-Ceyranbatan Arıtma Tesisi." 2015. http://www.tega.com.tr/baku-ceyranbatan-aritmatesisi/?lang=tr ; Azertag. "Prezident İlham Əliyev Ceyranbatan Ultrasüzgəcli Sutəmizləyici Qurğular Kompleksinin Açılışında Iştirak Edib." October 28, 2015. https://azertag.az/xeber/Prezident_Ilham_Aliyev_Ceyranbatan_Ultrasuzgecli_Sutemizleyici_Qurgular_Kompleksinin_achilisinda_istirak_edib_VIDEO-896553
174 Nigar Orujova, "World's Largest Surface Water Filtration Plant to Be Built in Baku." *Azernews*, February 20, 2013. Retrieved July 1, 2021, from, http://www.azernews.az/business/49963.html
175 Ibid.

176 Reliefweb, "FAO + Türkiye: Partnering for Food Security and Sustainable Management of Natural Resources." March 24, 2021. https://reliefweb.int/report/Türkiye/fao-Türkiye-partnering-food-security-and-sustainablemanagement-natural-resources
177 Ibid.
178 The Food and Agriculture Organization (FAO), "Developing Capacity for Strengthening Food Security and Nutrition." http://www.fao.org/in-action/fsn-caucasus-asia/overview/en/
179 Reliefweb, "Food Security and Vulnerability Assessment in Armenia, Round 2, February 2021." March 5, 2021. https://www.wfp.org/countries/armenia https://reliefweb.int/report/armenia/food-security-and-vulnerability-assessment-armenia-round-2-february-2021
180 Reliefweb, "Food Security and Vulnerability Assessment in Armenia, November 2020." November 23 2020. Retrieved June 29, 2021 from https://reliefweb.int/report/armenia/food-security-and-vulnerability-assessmentarmenia-november-2020
181 Ibid.
182 UNEnvironmentProgramme."GeorgiaPushestoBolsterItsFoodSecurity."June12,2020. https://www.unep.org/news-and-stories/story/georgia-pushes-bolster-its-food-security
183 Ibid.
184 United Nations Human Rights Office of the High Commission, "Statement by Ms. Hilal Elver, United Nations Special Rapporteur on the Right to Food on Her Visit to the Republic of Azerbaijan, 1–11 October 2019." October 11, 2019. https://www.ohchr.org/EN/NewsEvents/Pages/DisplayNews.aspx?NewsID=25122&LangID=E
185 The Food and Agriculture Organization(FAO). "Improving Livelihoods and Food Security for Farmers in Azerbaijan." http://www.fao.org/policy-support/tools-and-publications/resources-details/en/c/1195478/
186 Ibid.
187 Oliver Bullough, "Armenian-Azerbaijani Water Politics." October 21, 2010. http://www.globalpolicy.org/component/content/article/198-natural-resources/49597-armenian-azerbaijaniwater-politics.html
188 Ibid.
189 Paul Goble, "Azerbaijan Diplomatic Academy. Water as a Foreign Policy Challenge for Azerbaijan." October 1, 2011. http://biweekly.ada.edu.az/vol_4_no_19/Water_as_a_foreign_policy_challenge_for_Azerbaijan.htm
190 Water Politics, "Turkish Dam Would Impact Georgia." January 15, 2012. http://www.waterpolitics.com/2012/01/15/turkish-dam-would-impact-georgia/

References

Akanda, A., S. Freeman, and M. Placht. "The Tigres-Euphrates River Basin: Mediating a Path Towards Regional Water Stability". *The Fletcher School Journal for Issues Related to Southwest Asia and Islamic Civilization.* (Spring 2007).

Alekperov, Adishirin, Ruslan Ch. Agamirzayev, and Ramil A. Alekperov. "Geo Environmental Problems in Azerbaijan," in *Urban Groundwater Management and Sustainability*, edited by J. H. Tellam et al. Berlin: Springer, 2006, pp. 39–58.

Aliev, Amir S. "Problems and Prospects of Water Resources Management in the Azerbaijan Republic," in *Sustainable Use and Development of Watersheds*, edited by I.E. Gonenc et al. Berlin: Springer Science, 2008, pp. 51–59.

Baev, Pavel K., and Indra Overland. "The South Stream versus Nabucco Pipeline Race Geopolitical and Economic (Ir)Rationales and Political Stakes in Mega-Projects." *International Affairs* 86, no. 5 (2010); 1075–1090.

BP. "BP Statistical Review of World Energy." June, 2020. https://www.bp.com/content/dam/bp/business-sites/en/global/corporate/pdfs/energy-economics/statistical-review/bp-stats-review-2020-full-report.pdf

BP. "BP Statistical Review of World Energy." June, 2022. https://www.bp.com/content/dam/bp/business-sites/en/global/corporate/pdfs/energy-economics/statistical-review/bp-stats-review-2022-full-report.pdf

Bullough, Oliver. "Armenian-Azerbaijani Water Politics." *Global Policy Forum, Institute for War and Peace Reporting*. October 21, 2010.

CEE Bankwatch Network. "Environmental and Social Impacts of the Khudoni Hydropower Project." March, 2014.

Çevre ve Şehircilik Bakanlığı. *İklim Değişikliği Ulusal Eylem Planı*. Ankara, 2011.

Choucri, N. *Global Accord: Environmental Challenges and International Responses*. Cambridge, MA: MIT Press, 1993.

Critchley, H., and T. Terriff. "Environment and Security," in *Security Studies for the 1990s*, edited by R. Schultz, R. Godson, and T. Greenwood. Washington, DC: Brassey's, 1993.

Dalacoura, K. "Turkish Foreign Policy in the Middle East: Power Projection and Post-Ideological Politics." *International Affairs* 97, no. 4 (2021): 1125–1142. doi: 10.1093/ia/iiab082.

Dariali Energy. "Dariali Hydro Power Plant Construction and Operation Project: Environmental and Social Impact Assessment Report". (Tbilisi, 2014).

Dinar, Shlomi. "Water, Security, Conflict, and Cooperation." *SAIS Review* 22, no. 2 (Summer/Fall 2002): 229–253.

Ebel, Robert, and Rajan Menon. *Energy and Conflict in Central Asia and the Caucasus*. Lanham, MD: Rowman and Littlefield, 2000.

Energy Charter Secretariat. "Azerbaijan: In-Depth Review of the Investment Climate and Market Structure in the Energy Sector." Brussels, 2011.

Energy Information Administration (EIA). *Country Analysis Executive Summary: Azerbaijan*. 2021. https://www.eia.gov/international/analysis/country/AZE.

Enerji Piyasası Düzenleme Kurumu. "Doğalgaz Piyasası Sektör Raporu." Ankara, 2017, p. 5.

Erşen, Emre, and Mitat Çelikpala. "Türkiye and the Changing Energy Geopolitics of Eurasia." *Energy Policy* 128 (2019): 584–592.

European Commission. *Green Paper: Towards a European Strategy for the Security of Energy Supply*. Brussels, 2001.

Feshbach, Murray, and Alfred Friendly, Jr. *Ecocide in the USSR: Health and Nature Under Siege*. New York: Basic Books, 1992.

Gleick, P. *The World's Water: 1998–1999*. Washington, DC: Island Press, 1998.

Goltz, Thomas. *Azerbaijan Diary: A Rogue Reporter's Adventures in an Oil-Rich, War-Torn, Post-Soviet Republic*. Armonk, NY: M.E. Sharpe, 1998.

Green, C. "Energy Intensity: Complexities, Implications, Controversies." Energy Institute, University of Wisconsin-Madison, February 2011.

Hamso, Bjorn. "Georgia – Infrastructure Pre-Investment Facility: P098850 – Implementation Status Results Report : Sequence 10." Washington, DC: World Bank Group, 2011.

Harmancioglu, N., N. Alpaslan, and E. Boelee. *Irrigation, Health and Environment: A Review of Literature from Türkiye*. Colombo, Sri Lanka: International Water Management Institute, IWMI working paper 6, 2001, p. 21.

Homer-Dixon, T. "On the Threshold: Environmental Changes as Cause of Acute Conflict." *International Security* 16, no. 2 (1991): 76–116.

International Energy Agency, Türkiye 2021 Energy Policy Review. https://iea.blob.core.windows.net/assets/cc499a7b-b72a-466c-88de-d792a9daff44/Türkiye_2021_Energy_Policy_Review.pdf

International Finance Corporation (IFC). "Adjaristsqali Environmental and Social Review." Project number 33435. October 3, 2013.

Ismayilov, Murad, and Norman A. Graham (eds.), *Turkish-Azerbaijani Relations: One Nation-Two States?* New York: Routledge, 2016.

Karanfil, Fatih. "Enerji-Büyüme-Çevre : Türkiye Üçgenin Neresinde?" *Uluslararası İlişkiler Dergisi* 5, no. 10 (2009): 14.

Kaya, I. *The Euphrates-Tigris Basin: An Overview and Opportunities for Cooperation under International Law.* Canada: a paper presented at "Water Resources Outlook for the 21st Century: Conflicts and Opportunities." September 1–6, 1997.

Kuprashvili, Natia. "Dam Projects Face Resistance in Georgia." *IWPR.Net.* March 5, 2012.

Landau, Jacob M. *Pan-Turkism: From Irredentism to Cooperation.* Bloomington: Indiana University Press, 1995.

LeVine, S. *Putin's Labyrinth: Spies, Murder, and the Dark Heart of the New Russia.* New York: Random House Inc., 2008.

Lowi, M. *Water and Power: The Politics of a Scarce Resource in the Jordan River Basin.* New York: Cambridge University Press, 1993.

Mathews, J. "Redefining Security." *Foreign Affairs* 68, no. 2 (1989): 162–177.

Morgan, P.M. *International Security: Problems and Solutions.* Washington, DC: CQ Press, 2006.

Muftuler-Bac, Meltem. "The Future of Energy Security for Europe: Türkiye's Role as an Energy Corridor." *Middle Eastern Studies* 47, no. 2 (2011): 361–378.

Myers, N. "Environment and Security." *Foreign Policy* no. 74 (Spring 1989): 23–41.

Netherlands Commission for Environment Assessment. "Advisory Review of the Environmental and Social Impact Assessment of the Khudoni Hydropower Project". NCEA OS24-B017, Georgia. 2013.

Öniş, Ziya, and Şuhnaz Yılmaz. "Türkiye and Russia in a Shifting Global Order: Co-Operation, Conflict and Asymmetric Interdependence in a Turbulent Region." *Third World Quarterly* 37, no. 1 (January, 2016): 71–95.

Popescu, Nicu. "Europe's Unrecognized Neighbours: The EU in Abkhazia and South Ossetia." CEPS Working Documents, 2007.

Porter, G. "Post-Cold War Global Environment and Security." *The Fletcher Forum of World Affairs* 14, no. 2 (1990): 332–344.

Republic of Azerbaijan National Water Supply and Sanitation Project. "Environmental Impact Assessment for Hovsan Wastewater Treatment Plant Sea Outfall Construction". Seurca-ASPI. November, 2009.

Republic of Türkiye Ministry of Energy and Natural Resources. 2022. http://www.enerji.gov.tr/enUS/Pages/International

Samson, P., and B. Charrier. *International Freshwater Conflict: Issues and Prevention Strategies.* Geneva: Green Cross International, 1997).

Shubitidze, Ana. *Toward Sustainable Municipal Wastewater Treatment System in Tbilisi, Georgia.* University of Lund: Thesis for International Institute for Industrial Environmental Economics, 2006.

Siddi, M. "The Scramble for Energy Supplies to Southeastern Europe: The EU's Southern Gas Corridor, Russia's Pipelines and Türkiye's Role," in *Türkiye as an Energy Hub? Contributions on Türkiye's Role in EU Energy Supply.* Nomos, 2017. Baden-Baden

Socor, Vladimir. "Nabucco-West Selected for Caspian Gas Delivery to Central Europe." *Eurasia Daily Monitor* 9, no. 124 (2012).

Sönmez, B. *Türkiye'de Çorak Islahi Arastirmalari ve Tuzlu Topraklarin Yönetimi.* Ankara: Sulanan Alanlarda Tuzluluk Yönetimi Sempozyumu Bildiriler Kitabı, 2004.

Suhrke, A. "Environmental Change, Migration and Conflict: A Lethal Feedback Dynamic?," in *Managing Global Chaos: Sources of and Responses to International Conflict*, edited by C. Crocker, F. Hampson, and P. Aall. Washington, DC: United States Institute of Peace Press, 1996.

TEVEM. "Türkiye Enerji ve Enerji Verimliliği Çalışmaları Raporu". (2010).

Topaloglu, F. "Trend Detection of Stream Flow Variables in Türkiye." *Frenius Environmental Bulletin* 15, no. 6 (2006): 644–653.

Turkay, M., S. Yılmaz, and B.S. Akça. "Türkiye'nin Enerji Verimliliği Haritası ve Hedefler." *KÜTEM Report*. May, 2012.

Unver, O., and R.K. Gupta. "Waterpricing: Issues and Options in Türkiye." *Water Resources Development* 19, no. 2(2003): 311–330.

Yavuz, M. Hakan. *Nostalgia for the Empire: The Politics of Neo-Ottomanism*. Oxford: Oxford University Press, 2020.

Yazar, Y. "Türkiye'nin Enerji Durumu ve Geleceği." *SETA Analiz* 31 (2010): 1–24.

Yılmaz, Şuhnaz, and Sever-Mehmetoğlu, Duygu. "Linking Foreign Policy and Energy Security: An Asset or a Liability for Türkiye." *Uluslararası İlişkiler* 13, no. 52 (2016): 119.

8

CONCLUSION

Towards Peace, Prosperity, and Environmental Sustainability in Eurasia

S. Frederick Starr argued in June 2022 that "Greater Central Asia is reeling from the twin shocks of the US withdrawal from Afghanistan and Vladimir Putin's invasion of Ukraine….Russia's aggressive behavior in Ukraine suggests it could be poised to assert itself in Central Asia as well."[1] One could argue, as we have suggested in Chapter 5, that the course of the war after that point—through mid-2023—may lead to a somewhat different, cautious though still disturbing, assessment. Russia's influence in the region appears weaker, and the ties to China, Türkiye, and the West appear to be strengthening. Moreover, Starr draws crucial attention to the impact that the Taliban regime in Afghanistan has had on disrupting the progress that had been made on Central Asian trade and transportation networks to the wider world, especially to the South. He also delineated the prospects for a supportive role that the West might play to help stabilize the region and its inclination to rise to meet the serious economic and political challenges.

There are important opportunities for regional collaboration to this end and indeed to help remediate the environmental damage that has been done from the Soviet legacy of irrigation and unsustainable cropping policies to the growing impact of climate change. The Water-Energy-Food nexus in Central Asia is complex, but not without opportunities and prospects for adaptation and resilience. The Astana International Forum, June 8–9, 2023 featured numerous high level experts, policy makers, and heads of state including President Tokayev of Kazakhstan, President Japarov of Kyrgyzstan, Prime Minister Abdulla Nigmatovich Aripov of Uzbekistan, and Kristalina Georgieva, Managing Director of the IMF, who all spoke with some optimism about Central Asia's economic and sustainable development future. This is more likely if the independent national governments in the region are able to collaborate more effectively without external political interference and with external investment and trade liberalization, as several of the headliners at the Forum argued.

DOI: 10.4324/9781003373735-8

President Tokayev argued that climate change is already *the* existential threat to nature and civilization. This is especially evident in Central Asia, and he stressed that the new generation of leaders in the region needed to proceed with implementing the substantial reforms enacted on paper. In his view "…much has changed in the last two years" especially in Kazakhstan.[2] Representatives from the EBRD and several renewable energy developers described substantial plans to invest in the region, and indeed predicted that Kazakhstan could well become the sustainable energy hub of the region.[3]

Speaking to the World Economic Forum in December 2022, International Energy Agency chief Fatih Birol stressed that "It's a fallacy that the world must choose between energy security and climate action," as he warned against using the current crisis triggered by Russia's war on Ukraine as an excuse to intensify dependence on fossil fuels. On the contrary, he argued that it should strengthen the drive for more investment in clean energy—including renewables, electrification, and energy efficiency—to mitigate dependence on imported oil and gas.[4]

Overall, in this volume, we explored the complex interactions of the Water-Energy-Food nexus, climate change impacts, and regional geopolitics. Current trends increase the significance of the energy-water nexus and its food security implications particularly for Central Eurasia and its neighbors in a highly volatile regional context. While the region is bestowed with rich energy resources (albeit unevenly distributed), water scarcity and environmental degradation compounded by climate change emerge as key challenges. Moreover, Russia's war on Ukraine has not only raised the stakes for European energy security, but it also aggravated regional challenges regarding food security. Thus, we posited that the region is "running on empty" as it strives to cope with an increasingly disrupted global economy while searching for security, more autonomy, and broad-based prosperity. Hence, in line with Putnam's logic of two-level games, domestic and external linkages are particularly significant for decision makers while dealing with economic constraints, energy needs and environmental degradation on the domestic front, and geopolitical dynamics of a turbulent region in the international realm. In this context, interests and impact of pivotal states, competition among and complementarities of key players regarding water, energy and food security, and the broader economic and security implications at the local, national, regional, and international levels shape the intricate links between energy, environment, and geopolitics.

In defining the path towards a more sustainable future, the United Nations Sustainable Development Goals (SDGs) include Clean Water and Sanitation; Affordable and Clean Energy; Climate Action; Life Below Water; Life on Land; Peace, Justice and Strong Institutions and Partnership for the Goals, which are directly related with the themes extensively discussed in the preceding chapters.[5] Their prioritization in achieving a precarious balance in the Security-Energy-Environment (SEE) triangle would be a decisive factor for paving the way towards peace, prosperity, and environmental sustainability in the regional and global context.

Water

The disaster of the desiccation of the Aral Sea illustrates the extremes to which water scarcity together with unsustainable use for excessive irrigation for inappropriate agricultural development can mean for beleaguered populations. The number of desertification refugee fishers of the Aral City region of Kazakhstan who had to move to the Almaty/Kapchagay Reservoir and Lake Balkhash fisheries and aquaculture enterprises was substantial. This, along with the well-documented environmental health problems faced by the population that remained in the Aral region of both Kazakhstan and Uzbekistan, are testimony to the human impact of water scarcity. The remediation efforts by Kazakhstan to return water to the little Aral Sea and begin to restore habitat and commercial fishing do suggest that lessons have been learned by the independent Central Asia governments. Free from the Soviet rule, they can develop environmentally sustainable policies and programs, especially with the assistance of external investment and grants from Western and Asian governments, the World Bank and regional development banks like the EBRD and the ADB. Nonetheless, the challenges are immense. The dramatic amount of glacier melt in the mountains of China, Kyrgyzstan, and Tajikistan in recent years is well documented. But the impact of climate change on snow fields and permafrost areas must be considered as well. The Asian "water tower" is vulnerable, and the impact is not just to reduce flows downstream in the long term but also overflow the mountain lakes and lead to downstream flooding. Moreover, in the near term, there are already signs that downstream lakes and reservoirs (like Lake Balkhash and the Kapchagay Reservoir in Kazakhstan) are losing depth and breadth and may ultimately face the same fate as the Aral Sea if conservation and cross-border water sharing is not improved substantially.

The fragmentation of water resources and transboundary nature of river basins also constrain water security and escalate tensions among the riparian states. Globally, there are 276 transboundary basins, shared by 148 countries accounting for 60% of the global freshwater flow. In the South Caucasus, as discussed in Chapter 7, in the case of the Kura-Aras river basin, Georgia, Armenia, and Azerbaijan have failed to cooperate on the management of the Kura-Aras and often accuse each other for dumping pollutants into its waters. In the greater Middle East, disputes over the Tigris and Euphrates River basin also strain the ties between Türkiye, Syria, and Iraq regarding water quantity and quality.

Chapter 4 makes very clear that Central Asian states, while having learned some lessons the Aral Sea disaster, have not fully faced up to continuing threat of water scarcity and quality. For example, the Ili River-Lake Balkhash ecosystem in Kazakhstan (and China) is in serious danger of replicating the Aral Sea desiccation if conservation measures are not implemented in a cross-border water sharing framework.

In the regional context, there is a pressing need for better governance based on transparency, compliance with legal and regulatory frameworks, institutionalized mechanisms for data sharing and coordination, and inclusion of all stakeholders. In response to the mounting challenges at the regional and national level, it is critical to strengthen the infrastructure, institutions, incentives, and information systems that guide and support water management. The World Bank warns that "Water scarcity affects more than 40% of the global population. Water-related disasters account for 70% of all deaths caused by natural disasters."[6] Moreover, to feed 10 billion people by 2050, a 50% increase in agricultural production (which currently consumes approximately 70% of the resource), and a 15% increase in water withdrawals would be necessary.[7] Hence, building climate resilience and integrated water management is vital for sustainability and human security.

Energy

Jason Bordoff and Meghan O'Sullivan argued in 2023 that the expected "...geopolitical benefits of the coming transition to cleaner, greener energy....evaporated when Russia invaded Ukraine in 2022."[8] Bordoff and O'Sullivan made a persuasive case that the need to focus on energy security issues and policies is far from over despite the optimism building prior to 2022. This optimism had derived from the push for development of renewable energy sources, the commitment in much of the world to reducing consumption of coal, and the increased development of natural gas fields, including those accessed by fracking. The range of new fossil fuel suppliers seemed to reduce dependence on unreliable, politically minded suppliers (like Russia) and cartel price support operations (like OPEC). Clearly, the changing geopolitical environment "...can affect the pace and scope of the transition to clean energy. Before the Russian invasion of Ukraine, European countries and the United States seemed committed to transforming their economies to achieve net-zero carbon emissions in the coming decades."[9] Bordoff and O'Sullivan concluded:

> Policymakers must grasp the new risks to energy security and modernize their toolkits to combat them. Doing so is not a distraction from addressing climate change but central to it; without this shift, energy crises might derail the drive to net-zero emissions. In the not-so-distant past, officials and experts thought that excessive fears about energy security might hinder the fight for the climate. Today, the opposite is true: as the transition to a net-zero world proceeds, the bigger danger to the climate will be insufficient attention to energy security.[10]

Analysts and policy makers have begun to argue that the post-war reconstruction in Ukraine (when it finally begins) should focus on making it central to Europe's green transition. For example, Rostyslav Shurma, deputy head of the Ukrainian president's office in charge of economic policy argued: "Carbon-free energy should be a key target for investments in the country's reconstruction."[11] There is some evidence that this view is shared by Central Eurasian policy makers, even those with long records of commitment to fossil fuel development and exports like Azerbaijan and Kazakhstan. Diversification of the economies more generally has become a major focus, as Kazakhstan and Uzbekistan, for example, have addressed food security more seriously. Nearly all Central Eurasian countries have devoted new attention to renewable energy technologies, at least partly, because the threat of disruption to their fossil fuel systems and supply and export networks has seemed more real of late and the need to cut carbon emissions seems more dire.

Some of this attention remains focused on hydroelectric power generation systems. President Japarov, for example, in his speech to the Astana International Forum June 8, 2023, pointed to the need to develop small and medium hydroelectric plants on more of the 252 large and medium rivers in Kyrgyzstan. The Republic of Georgia has long sought to develop its bountiful river resources for hydroelectric generation. The Rogun Dam project in Tajikistan is obviously another important step in renewable energy development. Of course, there is the downside of potential for water flow reduction to downstream countries for agricultural irrigation and human consumption, as well as negative impacts on fish habitat and upstream spawning processes. It is possible to construct fish ladders around new dam installations, of course, as the elaborate structures operated in the Columbia River in Oregon illustrate—for example, the Bonneville Dam fishway established in the 1940s. Another exemplar is the famous Pitlochry fish ladder on the River Tummel, Scotland, established in 1951. On the negative side is the devastating impact that dam installations on the Volga River had in decimating the sturgeon population and beluga caviar production in the USSR. The massive dam established at Stalingrad/Volgograd in 1960–1961 is perhaps the most devastating example, with estimates suggesting that the sturgeon catch dropped by 80% as a consequence.[12] The Soviet disregard for environmental damage in its well-known tendency to celebrate construction that demonstrated man's mastery over nature with little concern for broader impacts in the water-energy-food nexus is clear in these projects. This was remarkable in the 1960s for a country that presumably had little real need for destructive hydroelectric generation, given its massive oil and gas reserves and burgeoning nuclear energy development progress.

However, the most sustainable immediate potential for expansion of renewable energy is in solar and wind technology, and it seems clear that Central Eurasian governments have begun to develop financing and network planning for extensive solar and wind fields. Obviously, there is no shortage of sun and wind in Armenia, Azerbaijan, Kazakhstan, Turkmenistan, and Uzbekistan where water resources are growing particularly scarce. Fortunately, the cost of solar panels and wind turbines has come down dramatically, and even the prospect for domestic production of the systems has begun to seem viable.

Food

Food security is an important challenge in much of Central Eurasia given the impact of climate change and regional instability. As indicated in Chapter 4, domestic agricultural development has evolved substantially in recent years in response to government reform efforts, private sector entrepreneurial activity, and efforts to adjust to water scarcity. But there remain many challenges.

Kazakhstan, for example, still imports about 40% of its food from Russia, at a time when Russia seems increasingly isolated and strained in its operations against Ukraine. President Tokayev's multivector foreign policy includes some effort to intensify trade relations with other countries in the region, but there is also more scope for expanding domestic food production through diversification away from cotton cultivation, enhancing aquaculture production and turning Kazakhstan's northern steppe grasslands into enhanced livestock production.[13]

Uzbekistan has been quite successful in increasing aquaculture production and has diversified its agriculture in important respects.[14] As noted in Chapter 4, cotton production has been reduced substantially, with reforms also to eliminate child labor in the harvesting. A variety of fruits and vegetables have been cultivated instead, with due attention to those that require less water than cotton. Efforts to provide farmers with modern technology and information on more sustainable techniques with less chemical pest control and fertilizer have been incorporated. The potential for growth in this modernization of farming is large.

The Republic of Georgia has been moderately successful in recovering its prominence as a key producer of fruit and vegetables during the Soviet period. Mikhail Saakashvili's focus on the IT revolution as the key ingredient to Georgia's economic future led to lost investment in agriculture and related loss in traditional expertise. The Agricultural University of Georgia has seen new prominence in supporting agricultural development, and government investment has returned to some extent. Partnerships with Western universities with strong agricultural technology programs have been useful, and agricultural trade with other Central Eurasian and European countries has grown. The export of the highly respected Georgian wines is a noteworthy success story, helping to reclaim Georgia's reputation as the historical birthplace of winemaking.

The Water-Energy-Food Nexus

The challenges facing Central Eurasia demonstrate the complexity and the dire nature of the Water-Energy-Food Nexus as a source of threats to security. Obviously, this is not unique to Central Eurasia. The United Nations Water Section argued in 2022 that: "Water, food and energy form a nexus at the heart of sustainable development. Agriculture is the largest consumer of the world's freshwater resources, and water is used to produce most forms of energy."[15] According to the UN, 72% of all water withdrawals are used by agriculture, 16% by municipalities for households and services, and 12% by industries.[16] The drier the geography, the more precious the sources of fresh water. Given that the water sources in much of Central Eurasia are transboundary in nature, and there are substantial differences in energy resources, the nexus is apparent, as discussed in Chapter 4. In Chapter 7, we also explored the implications of these interactions for the South Caucasus and also how they affected relations with Türkiye. The prospects for managing the nexus effectively have been limited since the dismantling of the Soviet centralized power which had been able to enforce distribution/sharing of resources across republic boundaries. There have been several attempts to recreate this sharing system through intergovernmental negotiations. The complementarities are there, perhaps now the political will is building.

There appears to be new commitment to resolving WEF tensions within Central Eurasia. The impact of climate change already felt through rising temperatures and glacial melt seems to have raised the urgency for cooperative action. And in transboundary settings with different endowments of water and oil or gas, there must be tradeoffs and efforts to increase resource use efficiency and take advantage of complementarities.[17] The political will and leadership is also extremely critical in addressing these complex challenges. This did not seem possible when Presidents Karimov and Nazarbayev were in power. It may be more likely now. Regional cooperation on serious challenges must be pursued vigorously. Domestic interests and inequalities must be incorporated into the deal making. Protests like those that turned violent and led to violent repression, with Russian help, in Kazakhstan in January 2022, illustrate that efforts to reduce poverty and promote broad-based prosperity must be part of the program.

Theoretical and Policy Implications: Climate Change and the WEF Nexus Meet Geopolitics in Central Eurasia

A careful review of Central Eurasian government documents and our detailed discussions with public officials and relevant scientists make it clear that there is serious concern about the ways in which climate change exacerbates the challenges of the Water-Energy-Food nexus in the region. Increased glacier melt in the mountains of China, Kyrgyzstan, and Tajikistan provide enhanced seasonal downstream flows in the Ili, Amu Darya, and Syr Darya river systems for the time being, but the prospective long-term effects of the melt are clear. Adjustments must be and to some extent have been made to human consumption patterns and agricultural production systems. Much remains to be done.

There is now a large literature that examines the water-energy-food nexus in the context of many countries and regions, largely with a focus on developing countries. A. P. Hejnowicz and a large interdisciplinary team of colleagues have reviewed much of this literature in detail. Generally, they found considerable progress in the application of the framework but stressed that most of the attention had been on scientific measurement and modeling of the interaction between the key variables, based on remote sensing data and field measurements of water levels, river flows, land use, and carbon emission levels. They called for an effort to bring economic, political, and social factors into the frameworks more explicitly.[18] With this in mind, we sought to bring these factors and the corresponding policy development efforts explicitly into our assessment of WEF in Central Eurasia.

Central Eurasian policy makers increasingly speak of the interactions between water, energy, and food security concerns. This is not surprising, of course, given the geography, topography, and worsening climate of the region, which are hard to ignore. The question that remains is whether the policy makers actually have the will and the power to implement demanding policies and programs that will address the complexities in concert with neighboring governments. Some of the WEF literature provides guidance on how to make the necessary tradeoffs and difficult sacrifices, and there is some evidence that recent negotiations and reforms under the auspices of the new Presidents in Kazakhstan, Kyrgyzstan, and Uzbekistan are taking WEF tradeoffs seriously.[19]

Environmental security appears to be rising as a key area for policy and assessment in Central Eurasia, as decision makers assess the broader geopolitical context that is rapidly changing and is already dramatically different from January 2022. What then is the emerging structure of the international system? How disrupted is it in terms of constraints on normal economic interactions and supply chains? This is critical to those closest to the disastrous military conflicts underway, but how far does the disruption extend? UN World Food Programme officials regularly warn of the impact of reduced grain, sunflower oil, and fertilizer shipments from Ukraine that normally would go to the Middle East and North Africa. This may be only the beginning of the detrimental impact that the war has had on food and energy security.

Is Russia's invasion of Ukraine evidence of a destabilized multipolar international system in which the rule of law and international norms, mostly in play for much of the decades since World War II, are no longer operating? Are international institutions like the World Bank, International Monetary Fund, and the World Trade Organization, not to mention the United Nations Security Council and the International Criminal Court, hopelessly broken, measurably in decline or increasingly irrelevant? Do the answers to these questions matter much to Central Eurasian governments and societies?

The structure and evolution of the international system has been a source of theoretical debate among International Relations scholars and practitioners since at least the Balance of Power system that emerged with and after the Napoleonic Wars.[20] The post-World War II bipolar system of two superpowers seemed at times to have some stability at least to the extent that major powers did not engage in full-scale conventional warfare, and nuclear deterrence seemed to be preserved with Mutually Assured Destruction (MAD). The end of the Cold War suggested to many, most notably US President George H. W. Bush, that a New World Order was now possible. The remnants of the Bretton Woods System of economically liberal trade and investment institutions established toward the end of World War II or shortly thereafter remained somewhat stable and constructive for most of the rest of the Twentieth Century.

This simplistic sketch of more than 50 years of world history is only useful to the extent to which we can draw comparisons to the system structure we now face. Richard Haass offered an important starting point with his 2017 book: *A World in*

Disarray? American Foreign Policy and the Crisis of the Old Order. Interestingly, even in early 2017 before the Trump Administration began to replace the previous US commitment to trade policies of enlightened self-interest with an America first posture, challenging liberal trade deals and the US role in NATO, Haass argued that the rules, policies, and institutions that had guided the world since World War II had largely run their course. Great-power rivalry was returning, and a World Order 2.0 needed to be constructed. He likened the period to the unsettling and protectionist period of the interwar period captured so well by English historian E.H. Carr in *The Twenty Years' Crisis: 1919–1939*.[21]

President Putin's annexation of Crimea in 2014, and especially his full-scale invasion of Ukraine in February 2022, reaffirmed the view that rules and policies had changed at least in his mind, and it was not clear that the post-World War II international institutions were up to the challenge. NATO may have been the exception with the Scandinavian enlargement and the massive and unified security assistance underway—most likely an unintended consequence of Putin's risk taking. But the common mantra for at least the last few years was that the international system had been disrupted by this brutal war, on top of the COVID-19 pandemic and the associated rise of protectionism amid global supply chain disruptions. Globalization was under attack, and analysts and policy makers, especially Putin and Xi Jinping, began to speak of the expected evolution to a multipolar international system. Clearly, these last two points will remain popular topics for discussion, but the alternative of reforming (democratizing?) globalization policies and instruments seriously, as Joseph E. Stiglitz suggests,[22] may be a better solution to the first, and the notion of rising multipolarity may be overstated or at least premature to describe the international system at this stage, as Brooks and Wohlforth argued in May 2023.[23] While there is an extensive debate among scholars and policymakers about the disruptions and transformation in the international system,[24] there is lack of a consensus regarding how to characterize the current international system.

Brooks and Wohlforth acknowledged that Russia still has claim to a very large nuclear weapons force, and China's remarkable economic rise to global manufacturing power is clear. However, neither country comes close to match the United States in the full range of ingredients of great power status. Beyond Russia and China, there are even less likely contenders for great power status anytime soon. In the words of Brooks and Wohlforth:

> The world is neither bipolar nor multipolar and it is not about to become either. Yes, the United States has become less dominant over the past 20 years, but it remains at the top of the global power hierarchy—safely above China and far, far above every other country....For the system to be multipolar, however, its workings must be shaped largely by the three or more roughly matched states at the top. The United States and China are undoubtedly the two most powerful countries, but at least one more country must be roughly in their league for multipolarity to exist. This is where claims of multipolarity fall apart. Every

> country that could plausibly rank third—France, Germany, India, Japan, Russia, the United Kingdom—is in no way a rough peer of the United States or China.[25]

This power imbalance is illustrated by the success of US President Biden to maximize the collective action of a very large system of allies in NATO, Asia, and Oceana to construct the largest effort ever of global economic and political sanctions levied against Russia discussed in Chapter 3. The Russian invasion of Ukraine did have an initial negative impact on global energy security and progress toward reducing climate carbon emissions by reducing coal consumption, but the United States and its allies have been able to ramp up the development of renewable energy sources to a remarkable degree and aim to reduce dramatically Europe's dependence on Russia for oil and gas, as discussed in Chapters 2 and 3.

There is considerable talk and indeed some concerted action to deglobalize the international economic system, given the vulnerabilities revealed by the Russian invasion, the COVID-19 pandemic, and worrying financial instability, as well as the concern about technology transfer through foreign direct investment and open trade of high technology systems. President Biden's "Chips Bill" initiative and related efforts to respond to this is important, though somewhat worrying to major US trading partners who fear a more general slip into global protectionism and nationalist industrial targeting.[26]

The Director-General of the World Trade Organization, Ngozi Okonjo-Iweala of Nigeria argued in September 2022:

> The current re-think on supply chain resilience in the wake of the COVID-19 pandemic and the war in Ukraine creates a window of opportunity to bring these poor countries from the margins of the global economy to the mainstream.... Expanding this process of what I call 're-globalization' would accomplish objectives that 'de-globalization' cannot.[27]

Joseph E. Stiglitz did not label it re-globalization, but as early as 2018, he argued that there are legitimate grievances against globalization and that reform was necessary and feasible:

> But I argued that the central problem with globalization was that it reflected an agenda driven by multinational corporations....America is one of the few countries which, on its own, can try to change the rules of the game. I have explained why protectionism won't work for the United States. Other countries which go down the protectionist route on their own are likely to fare even worse....But there are alternative ways of reforming globalization which hold out the promise of improving the lot of all, or at least most, citizens.[28]

The relevance of these musings to the conclusion of this volume is simply that the governments of Central Eurasia are in an unusual geopolitical situation. No matter how much consensus is reached ultimately among international system theorists and critics of the weakening of international norms and institutions, there may now be opportunities for closer regional cooperation without undue influence or indeed control by extra-regional powers. The environmental and economic development challenges are immense, and external investment and liberal trade policies will be necessary. But an aggressive effort to develop sustainable water, energy, and food security policies for *national security* will pay dividends well into the future, and there may well be valuable energy export growth to be pursued in the meantime, given the sanction regime impacts on Russian exports.

How can we best visualize the geopolitical evolution of Eurasia? Timothy Garton Ash touches on this in his 2023 thought-provoking article: "Postimperial Empire—How the War in Ukraine is Transforming Europe."[29] One can easily extend some of his key points to show relevance to the current situation facing the independent states of Central Eurasia and the extent to which his final sentence has relevance beyond Europe proper:

> Here, then, is the surprising prospect that the war in Ukraine reveals: The EU as a postimperial empire, in strategic partnership with an American postimperial empire, to prevent the comeback of a declining Russian empire and constrain a rising Chinese one.[30]

Ash quotes Dmytro Kuleba, Ukraine's foreign minister as describing the European Union as "the first ever attempt to build a liberal empire," in contrast to Russia's use of military force to regain a more traditional colonial empire.[31]

As discussed in Chapters 5 and 6, while Russia's sphere of influence has been waning in Central Eurasia, in part due to China's Belt and Road Initiative, the Chinese presence in the region is rapidly expanding. However, the Russian invasion of Ukraine has also reinforced existing sentiment among Central Eurasian leaders and populations for "multivector international relations" including with EU countries, Türkiye, and the United States. This all has significance for the WEF nexus in the region. Ultimately, the interaction of local needs, national priorities, and international dynamics will determine how energy, environment, and geopolitics shape Eurasia's future in a world of drastic uncertainty and change.

Notes

1 S. Frederick Starr, *Rethinking Greater Central Asia: American and Western Stakes in the Region and How to Advance Them* (Washington, DC: Central Asia-Caucasus Institute-Silk Road Paper, June 2022), p. 4.
2 Astana International Forum, Speech to the Plenary session, June 8, 2023.
3 Astana International Forum. Panel on renewable energy prospects, June 8, 2023.
4 World Economic Forum, "Why Climate and Energy Security are Closely Intertwined?" December 8, 2022. https://www.weforum.org/agenda/2022/12/why-climate-and-energy-security-are-closely-intertwined
5 United Nations, *Sustainable Development: The 17 Goals*. https://sdgs.un.org/goals
6 The World Bank, *Water Resources Management*. https://www.worldbank.org/en/topic/waterresourcesmanagement#:~:text=Water%20Resources%20Management%20(WRM)%20is,support%20and%20guide%20water%20management
7 Ibid.

8 Jason Bordoff and Meghan L. O'Sullivan, "The Age of Energy Insecurity: How the Fight for Resources Is Upending Geopolitics." *Foreign Affairs*, May/June, 2023, pp. 104–105.
9 Ibid., p. 107.
10 Ibid., p. 119.
11 As quoted in Martin Sandbu, "Ukraine's Rapid Reconstruction Will Be in Europe's Own Interest." *Financial Times*, June 4, 2023.
12 Janet M. Hartley, *The Volga: A History of Russia's Greatest River* (New Haven, CT: Yale University Press, 2021), p. 312.
13 Norman A. Graham, et al., "The Dynamics of Kazakhstan's Fisheries Sector: Prospects for Expansion," and Steven G. Pueppke, "Kazakhstan's Northern Steppe Grasslands: An Underutilized Resource to Increase Livestock Production," in *The New Kazakhstan: Steering the Steppe Amid a Global Storm*, ed. Marc Uzan (Paris: Reinventing Bretton Woods Committee, 2023).
14 Norman A. Graham, Stephen G. Pueppke, Sabir Nurtazin, Talgarbay Konysbayev, Farid Gibadulin, and Meiirli Sailauov, "The Changing Dynamics of Kazakhstan's Fisheries Sector: From the Early Soviet Era to the Twenty-First Century." *Water* (May, 2022); and Norman A. Graham, Steven G. Pueppke, and Talgat Uderbayev, "The Current Status and Future of Central Asia's Fish and Fisheries: Confronting a Wicked Problem," *Water* (April, 2017).
15 United Nations Water, *Water, Food and Energy*, 2021. https://unwater.org/water-facts/water-food-and-energy/
16 Ibid.
17 United Nations Economic Commission for Europe, *Water-Food-Energy-Ecosystem Nexus: Why a "Nexus" Approach to Transboundary Cooperation?* UNECE(/) Environmental Policy (/NODE/4) Water(/NODE/23), December 22, 2022.
18 A.P. Hejnowicz, J.P.R. Thorn, M.E.Giraudo, J.B. Sallach, S.E. Hartley, J. Grugel, S.G. Pueppke, and L. Emberson, Appraising the Water-Energy-Food Nexus from a Sustainable Development Perspective: A Maturing Paradigm?" *Earth's Future*, October, 2022, pp. 1–34.
19 S.G. Pueppke, S.T. Nurtazin, N.A. Graham, and J. Qi, "Central Asia's Ili River Ecosystem as a Wicked Problem: Unraveling Complex Interrelationships at the Interface of the Water, Energy, Food." *Water*, 10 no. 5 (2021).
20 See: Edward Vose Gulick, *Europe's Classical Balance of Power: A Case History of the Theory and Practice of One of the Great Concepts of European Statecraft* (New York: W.W. Norton & Co., 1955); and F.H. Hinsley, *Power and the Pursuit of Peace: Theory and Practice in the History of Relations between States* (Cambridge, UK: Cambridge University Press, 1963), pp. 153–271; and Paul Kennedy, *The Rise and Fall of the Great Powers* (New York: Random House, 1987), espec. Ch. 4–8.
21 Richard Haass, *A World in Disarray: American Foreign Policy and the Crisis of the Old Order*.
22 Joseph E. Stiglitz, *People, Power, and Profits: Progressive Capitalism for an Age of Discontent* (NY: W.W. Norton & Co., 2019), especially Ch. 3–5; and Joseph E. Stiglitz, *Globalization and Its Discontents—Revisited.*
23 Stephen G. Brooks and William C. Wohlforth, "The Myth of Multipolarity: American Power's Staying Power." *Foreign Affairs*, May/June 2023.
24 Alexander Cooley and Daniel Nexon, Exit from Hegemony: Unraveling of the American Global Order, Oxford University Press, 2020; Joseph Borrell, High Representative of the European Union for Foreign Affairs and Security, "How to Revive Multilateralism in Multipolar World?", The European Union External Action, March 16, 2021, https://www.eeas.europa.eu/eeas/how-revive-multilateralism-multipolar-world_en
25 Ibid., pp. 78–79.
26 See: Chris Miller, *Chip War: The Fight for the World's Most Critical Technology* (New York: Simon & Schuster, Inc., 2022).
27 Ngozi Okonjo-Iweala, delivering the Rafael M. Salas Memorial Lecture sponsored by the United Nations Fund for Population Activities, New York, September 21, 2022.

28 Stiglitz, Globalization and Its Discontents—Revisited, p. 74.
29 Timothy Garton Ash, "Postimperial Empire: How the War in Ukraine is Transforming Europe," *Foreign Affairs*, May/June 2023, pp. 64–75.
30 Ibid., p. 75.
31 Ibid., p. 72.

References

Ash, Timothy Garton. *Homelands: A Personal History of Europe.* New Haven, CT: Yale University Press, 2023.

Ash, Timothy Garton. "Postimperial Empire: How the War in Ukraine Is Transforming Europe." *Foreign Affairs* (May/June, 2023): 64–75.

Bordoff, Jason, and Meghan L. O'Sullivan. "The Age of Energy Insecurity: How the Fight for Resources Is Upending Geopolitics." *Foreign Affairs* (May/June, 2023): 104–119.

Brooks, Stephen G., and William C. Wohlforth. "The Myth of Multipolarity: American Power's Staying Power." *Foreign Affairs* (May/June, 2023): 76–91.

Graham, Norman A., Steven G. Pueppke, Talgat Uderbayev. "The Current Status and Future of Central Asia's Fish and Fisheries: Confronting a Wicked Problem." *Water* (April 2017): 1–16, 1–22.

Graham, Norman A., Stephen G. Pueppke, Sabir Nurtazin, Talgarbay Konysbayev, Farid Gibadulin, and Meiirli Sailauov. "The Changing Dynamics of Kazakhstan's Fisheries Sector: From the Early Soviet Era to the Twenty-First Century." *Water* (May 2022).

Graham, Norman A., et al. "The Dynamics of Kazakhstan's Fisheries Sector: Prospects for Expansion," in *The New Kazakhstan: Steering the Steppe Amid a Global Storm*, edited by Marc Uzan. Paris: Reinventing Bretton Woods Committee, 2023.

Gulick, Edward Vose. *Europe's Classical Balance of Power: A Case History of the Theory and Practice of One of the Great Concepts of European Statecraft.* New York: W.W. Norton & Co., 1955.

Hejnowicz, A.P., J.P.R. Thorn, M.E.Giraudo, J.B. Sallach, S.E. Hartley, J. Grugel, S.G. Pueppke, and L. Emberson. "Appraising the Water-Energy-Food Nexus From a Sustainable Development Perspective: A Maturing Paradigm?" *Earth's Future* 10, no. 12 (October, 2022): 1–34.

Hinsley, F.H. *Power and the Pursuit of Peace: Theory and Practice in the History of Relations between States.* Cambridge, UK: Cambridge University, 1963.

Kennedy, Paul. *The Rise and Fall of the Great Powers: Economic Change and Military Conflict from 1500 to 2000.* New York: Random House, 1987.

Miller, Chris. *Chip War: The Fight for the World's Most Critical Technology.* New York: Simon & Schuster, Inc., 2022.

Pueppke, S.G., S.T. Nurtazin, N.A. Graham, and J. Qi. "Central Asia's Ili River Ecosystem as a Wicked Problem: Unraveling Complex Interrelationships at the Interface of the Water, Energy, Food." *Water* 10, no. 5 (2021): 1–21.

Pueppke, Steven G. "Kazakhstan's Northern Steppe Grasslands: An Underutilized Resource to Increase Livestock Production," in *The New Kazakhstan: Steering the Steppe Amid a Global Storm*, edited by Marc Uzan. Paris: Reinventing Bretton Woods Committee, 2023.

Starr, S. Frederick. *Rethinking Greater Central Asia: American and Western Stakes in the Region and How to Advance Them.* Washington, DC: Central Asia-Caucasus Institute-Silk Road Paper. June, 2022.

Uzan, Marc. *The New Kazakhstan: Steering the Steppe Amid a Global Storm*, edited by Marc Uzan. Paris: Reinventing Bretton Woods Committee, 2023.

INDEX

Note: **Bold** page numbers refer to tables; *italic* page numbers refer to figures and page numbers followed by "n" denote endnotes.

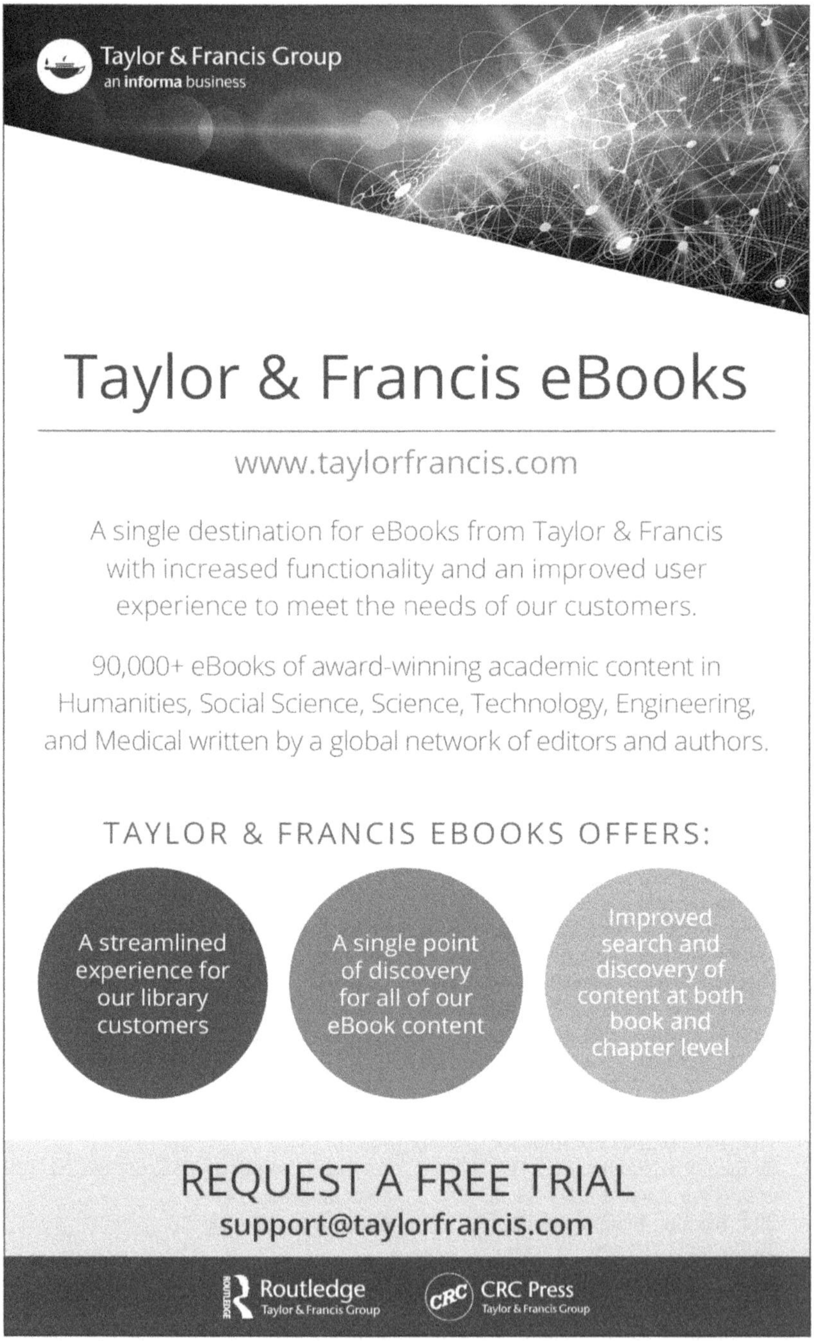
Taylor & Francis Group
an informa business
Taylor & Francis eBooks
www.taylorfrancis.com
A single destination for eBooks from Taylor & Francis with increased functionality and an improved user experience to meet the needs of our customers.
90,000+ eBooks of award-winning academic content in Humanities, Social Science, Science, Technology, Engineering, and Medical written by a global network of editors and authors.
TAYLOR & FRANCIS EBOOKS OFFERS:
A streamlined experience for our library customers
A single point of discovery for all of our eBook content
Improved search and discovery of content at both book and chapter level
REQUEST A FREE TRIAL
support@taylorfrancis.com
Routledge
Taylor & Francis Group
CRC Press
Taylor & Francis Group

For Product Safety Concerns and Information please contact our EU
representative GPSR@taylorandfrancis.com
Taylor & Francis Verlag GmbH, Kaufingerstraße 24, 80331 München, Germany

www.ingramcontent.com/pod-product-compliance
Ingram Content Group UK Ltd.
Pitfield, Milton Keynes, MK11 3LW, UK
UKHW020142200926
475702UK00005B/861
9781032447490